THE IMPACT OF
GENETICALLY ENGINEERED CROPS
ON FARM SUSTAINABILITY
IN THE UNITED STATES

D0957152

Committee on the Impact of Biotechnology on
Farm-Level Economics and Sustainability

Board on Agriculture and Natural Resources

Division on Earth and Life Studies

NATIONAL RESEARCH COUNCIL
OF THE NATIONAL ACADEMIES

THE NATIONAL ACADEMIES PRESS
Washington, D.C.
www.nap.edu

THE NATIONAL ACADEMIES PRESS   500 Fifth Street, N.W.   Washington, DC 20001

NOTICE: The project that is the subject of this report was approved by the Governing Board of the National Research Council, whose members are drawn from the councils of the National Academy of Sciences, the National Academy of Engineering, and the Institute of Medicine. The members of the committee responsible for the report were chosen for their special competences and with regard for appropriate balance.

This study was funded by the National Academies. Any opinions, findings, conclusions, or recommendations expressed in this publication are those of the authors and do not necessarily reflect the views of the organizations or agencies that provided support for the project.

International Standard Book Number-13:   978-0-309-14708-8 (Book)
International Standard Book Number-10:   0-309-14708-5 (Book)
International Standard Book Number-13:   978-0-309-14709-5 (PDF)
International Standard Book Number-10:   0-309-14709-3 (PDF)
Library of Congress Control Number: 2010927922

Additional copies of this report are available from the National Academies Press, 500 Fifth Street, NW, Lockbox 285, Washington, DC 20055; (800) 624-6242 or (202) 334-3313 (in the Washington metropolitan area); Internet, http://www.nap.edu.

Suggested Citation: National Research Council. 2010. The Impact of Genetically Engineered Crops on Farm Sustainability in the United States. Washington, DC: National Academies Press.

THE NATIONAL ACADEMIES
Advisers to the Nation on Science, Engineering, and Medicine

The **National Academy of Sciences** is a private, nonprofit, self-perpetuating society of distinguished scholars engaged in scientific and engineering research, dedicated to the furtherance of science and technology and to their use for the general welfare. Upon the authority of the charter granted to it by the Congress in 1863, the Academy has a mandate that requires it to advise the federal government on scientific and technical matters. Dr. Ralph J. Cicerone is president of the National Academy of Sciences.

The **National Academy of Engineering** was established in 1964, under the charter of the National Academy of Sciences, as a parallel organization of outstanding engineers. It is autonomous in its administration and in the selection of its members, sharing with the National Academy of Sciences the responsibility for advising the federal government. The National Academy of Engineering also sponsors engineering programs aimed at meeting national needs, encourages education and research, and recognizes the superior achievements of engineers. Dr. Charles M. Vest is president of the National Academy of Engineering.

The **Institute of Medicine** was established in 1970 by the National Academy of Sciences to secure the services of eminent members of appropriate professions in the examination of policy matters pertaining to the health of the public. The Institute acts under the responsibility given to the National Academy of Sciences by its congressional charter to be an adviser to the federal government and, upon its own initiative, to identify issues of medical care, research, and education. Dr. Harvey V. Fineberg is president of the Institute of Medicine.

The **National Research Council** was organized by the National Academy of Sciences in 1916 to associate the broad community of science and technology with the Academy's purposes of furthering knowledge and advising the federal government. Functioning in accordance with general policies determined by the Academy, the Council has become the principal operating agency of both the National Academy of Sciences and the National Academy of Engineering in providing services to the government, the public, and the scientific and engineering communities. The Council is administered jointly by both Academies and the Institute of Medicine. Dr. Ralph J. Cicerone and Dr. Charles M. Vest are chair and vice chair, respectively, of the National Research Council.

www.national-academies.org

---

[1]The views expressed here are those of the authors and may not be attributed to the Economic Research Service or the U.S. Department of Agriculture.

Preface

Not since the introduction of hybrid corn seed have we witnessed such a sweeping technological change in U.S. agriculture. Hundreds of thousands of farmers have adopted the first generation of genetically engineered (GE) crops since their commercialization in 1996. Although not all GE varieties that have been commercialized have succeeded, those targeted at improved pest control now cover over 80 percent of the acres planted to soybean, cotton, and corn—that is, almost half of U.S. cropland. Forecasts suggest an expansion in GE-crop plantings in many other countries.

GE crops originate in advances in molecular and cellular biology that enable scientists to introduce desirable traits from other species into crop plants or to alter crop plants' genomes internally. Those powerful scientific techniques have dramatically expanded the boundaries that have constrained traditional plant breeding. A new technology adopted so widely and rapidly has substantial economic, social, and environmental impacts on farms and their operators. Inevitably, both advantages and risks or losses emerge from such massive changes. The National Research Council has conducted multiple studies of specific aspects of GE crops, such as regulatory-system adequacy and food safety. However, the assigned tasks restricted the scope of their reports. As pressure mounts to expand the use of GE crops for energy, food security, environmental improvement, and other purposes, the scope and intensity of impacts will grow. Now is an opportune time to take a comprehensive look at the track record of GE crops and to identify the opportunities and challenges looming on the horizon. The National Research Council therefore supported the Committee on the Impact of Biotechnology on Farm-Level Economics and Sustainability to investigate this topic.

Despite the rapid spread of GE crops in U.S. agriculture, the technology continues to stir controversy around scientific issues and ideological viewpoints. The committee focused on the scientific questions associated with the farm-level impacts of the adoption of genetic-engineering technology and refrained from analyzing ideological positions, either pro or con. The committee adopted an "evidentiary" standard of using peer-reviewed literature on which to form our conclusions and recommendations. It is my hope that the report will give readers a firm grasp of the state of evidence or lack thereof on the scientific issues.

True to its charge, the committee adopted a sustainability framework that required an evaluation of environmental, economic, and social impacts of GE crops. Those three dimensions constitute the essential pillars of sustainability science. The summary and opening and closing chapters bring together the three perspectives for a fuller view of the technology's impact.

Given the controversies, readers will want to know the committee's composition and how it conducted its work in arriving at conclusions and recommendations. The biographies in Appendix C show a group of highly accomplished natural and social scientists who possess a broad array of research experience and perspectives on GE crops. That diversity of disciplines and expertise proved beneficial in introducing checks and balances in evaluating information from many angles. The committee members divided into teams to work on the various sections of the report on the basis of the members' expertise. The drafts by each team were reviewed by the full committee to ensure that everyone had a chance to comment on and improve and approve each section. I was continually impressed with the members' dedication to a hard-nosed and impartial evaluation of the best science on GE crops. Equally important, they kept open minds in considering new evidence presented by their colleagues and external experts. The result was a model multidisciplinary research process in which each of us learned from the others and improved the report quality.

In closing, I want to express my deep appreciation to the committee members for their tireless work and good humor in completing such a challenging task while working full time at their regular jobs. Their commitment and professionalism exemplify the best of public science. Each member made significant contributions to the final report. The committee also benefited from the testimony of several experts in the field and from the numerous comments of many conscientious external reviewers.

Finally, the quality of the report would not have been attained without excellent support and substantive input by study director Kara Laney, the valuable assistance of Kamweti Mutu, the insightful counsel of Robin Schoen, and the editorial work of the National Research Council.

David E. Ervin, *Chair*
Committee on the Impact of Biotechnology
on Farm-Level Economics and Sustainability

Acknowledgments

This report has been reviewed in draft form by persons chosen for their diverse perspectives and technical expertise in accordance with procedures approved by the National Research Council Report Review Committee. The purpose of the independent review is to provide candid and critical comments that will assist the institution in making its published report as sound as possible and to ensure that the report meets institutional standards of objectivity, evidence, and responsiveness to the study charge. The review comments and draft manuscript remain confidential to protect the integrity of the deliberative process. We wish to thank the following individuals for their review of the report:

David A. Andow, University of Minnesota, St. Paul
Charles M. Benbrook, The Organic Center, Enterprise, Oregon
Lawrence Busch, Michigan State University, East Lansing
Stephen O. Duke, Agricultural Research Service, U.S. Department of
    Agriculture, University, Mississippi
Robert T. Fraley, Monsanto Company, St. Louis, Missouri
Dermot J. Hayes, Iowa State University, Ames
Molly Jahn, University of Wisconsin, Madison
Nicholas Kalaitzandonakes, University of Missouri, Columbia
Peter M. Kareiva, The Nature Conservancy, Seattle, Washington
Michelle A. Marvier, Santa Clara University, California
Paul D. Mitchell, University of Wisconsin, Madison

George E. Seidel, Colorado State University, Fort Collins
Greg Traxler, The Bill & Melinda Gates Foundation, Seattle,
     Washington

Although the reviewers listed above have provided many constructive comments and suggestions, they were not asked to endorse the conclusions or recommendations, nor did they see the final draft of the report before its release. The review of the report was overseen by Drs. Alan G. McHughen, University of California, Riverside, and May R. Berenbaum, University of Illinois, Urbana-Champaign. Appointed by the National Research Council, they were responsible for making certain that an independent examination of the report was carried out in accordance with institutional procedures and that all review comments were carefully considered. Responsibility for the final content of the report rests with the authoring committee and the institution.

Contents

xiii

List of Tables, Figures, and Boxes

TABLES

xv

FIGURES

BOXES

Abbreviations and Acronyms

ACCase	acetyl-CoA carboxylase
ALS	acetolactate synthase
AMPA	aminomethylphosphonic acid
APHIS	Animal and Plant Health Inspection Service (U.S. Department of Agriculture)
BST	bovine somatotropin
Bt	*Bacillus thuringiensis*
Cry	crystal-like (protein)
DNA	deoxyribonucleic acid
EIS	environmental impact statement
EPA	U.S. Environmental Protection Agency
EPSPS	enzyme 5-enolpyruvyl-shikimate-3-phosphate synthase
GE	genetically engineered
GMO	genetically modified organism
HPPD	hydroxyphenylpyruvate dioxygenase
HR	herbicide-resistant
IPR	intellectual-property rights

IR	insect-resistant
ISHRW	International Survey of Herbicide Resistant Weeds
MCL	maximum contaminant level
NOP	National Organic Program
NOSB	National Organic Standards Board
OFPA	Organic Foods Production Act
PTO	U.S. Patent and Trademark Office
R&D	research and development
USDA	U.S. Department of Agriculture
USDA-ERS	U.S. Department of Agriculture, Economic Research Service
USDA-NASS	U.S. Department of Agriculture, National Agricultural Statistics Service
VR	virus-resistant

Summary

With the advent of genetic-engineering technology in agriculture, the science of crop improvement has evolved into a new realm. Advances in molecular and cellular biology now allow scientists to introduce desirable traits from other species into crop plants. The ability to transfer genes between species is a leap beyond crop improvement through previous plant-breeding techniques, whereby desired traits could be transferred only between related types of plants. The most commonly introduced genetically engineered (GE) traits allow plants either to produce their own insecticide, so that the yield lost to insect feeding is reduced, or to resist herbicides, so that herbicides can be used to kill a broad spectrum of weeds without harming crops. Those traits have been incorporated into most varieties of soybean, corn, and cotton grown in the United States.

Since their introduction in 1996, the use of GE crops in the United States has grown rapidly and accounted for over 80 percent of soybean, corn, and cotton acreage in the United States in 2009. Several National Research Council reports have addressed the effects of GE crops on the environment and on human health.[1] However, the effects of agricultural biotechnology at the farm level—that is, from the point of view of

---

[1]*Genetically Engineered Organisms, Wildlife, and Habitat: A Workshop Summary* (2008); *Safety of Genetically Engineered Foods: Approaches to Assessing Unintended Health Effects* (2004); *Environmental Effects of Transgenic Plants: The Scope and Adequacy of Regulation* (2002); *Ecological Monitoring of Genetically Modified Crops: A Workshop Summary* (2001); *Genetically Modified Pest-Protected Plants: Science and Regulation* (2000).

1

the farmer—have received much less attention. To fill that information gap, the National Research Council initiated a study, supported by its own funds, of how GE crops have affected U.S. farmers—their incomes, agronomic practices, production decisions, environmental resources, and personal well-being. This report of the study's findings expands the perspectives from which genetic-engineering technology has been examined previously. It provides the first comprehensive assessment of the effects of GE-crop adoption on farm sustainability in the United States (Box S-1).

In interpreting its task, the committee chose to analyze the effects of GE crops on farm-level sustainability in terms of environmental, economic, and social effects. To capture the broad array of potential effects, the committee interpreted "farm level" as applying both to farmers who do not produce GE crops and those who do because genetic engineering is a technology of extensive scope, and its influences on farming practices have affected both types of farmers. Therefore, to the extent that peer-reviewed literature is available, the report draws conclusions about the environmental, economic, and social effects, both favorable and unfavorable, associated with the use of GE crops for all farmers in the United States over the last 14 years. The report encapsulates what is known about the effects of GE crops on farm sustainability and identifies where more

---

BOX S-1
Statement of Task

An NRC committee will study the farm-level impacts of biotechnology, including the economics of adopting genetically engineered crops, changes in producer decision making and agronomic practices, and farm sustainability.

The study will:

- review and analyze the published literature on the impact of GE crops on the productivity and economics of farms in the United States;
- examine evidence for changes in agronomic practices and inputs, such as pesticide and herbicide use and soil and water management regimes;
- evaluate producer decision making with regard to the adoption of GE crops.

In a consensus report, the committee will present the findings of its study and identify future applications of plant and animal biotechnology that are likely to affect agricultural producers' decision making in the future.

research is needed. A full sustainability assessment of GE crops remains an ongoing task because of information gaps on certain environmental, economic, and social impacts.

Genetic-engineering technology continues to stir controversy around scientific issues and ideological viewpoints. This report addresses just the scientific questions and adopts an "evidentiary" standard of using peer-reviewed literature to form conclusions and recommendations. GE-trait developments may or may not turn out to be a cost-effective approach to addressing challenges confronting agriculture, but a review of their impact and an exploration of what is possible are necessary to evaluate their relative efficacy. Therefore, the report details the challenges and opportunities for future GE crops and offers recommendations on how crop-management practices and future research and development efforts can help to realize the full potential offered by genetic engineering.

KEY FINDINGS

The order of findings in this summary reflects the structure of the report and does not connote any conclusions on the part of the committee regarding the relative strength or importance of the findings. In general, the committee finds that genetic-engineering technology has produced substantial net environmental and economic benefits to U.S. farmers compared with non-GE crops in conventional agriculture. However, the benefits have not been universal; some may decline over time; and the potential benefits and risks associated with the future development of the technology are likely to become more numerous as it is applied to a greater variety of crops. The social effects of agricultural biotechnology have largely been unexplored, in part because of an absence of support for research on them.

Environmental Effects

Generally, GE crops have had fewer adverse effects on the environment than non-GE crops produced conventionally. The use of pesticides with toxicity to nontarget organisms or with greater persistence in soil and waterways has typically been lower in GE fields than in non-GE, nonorganic fields. However, farmer practices may be reducing the utility of some GE traits as pest-management tools and increasing the likelihood of a return to more environmentally damaging practices.

Finding 1. When adopting GE herbicide-resistant (HR) crops, farmers mainly substituted the herbicide glyphosate for more toxic herbicides.

However, the predominant reliance on glyphosate is now reducing the effectiveness of this weed-management tool.

Glyphosate kills most plants without substantial adverse effects on animals or on soil and water quality, unlike other classes of herbicides. It is also the herbicide to which most HR crops are resistant. After the commercialization of HR crops, farmers replaced many other herbicides with glyphosate applications after crops emerged from the soil (Figures S-1, S-2, and S-3). However, the increased reliance on glyphosate after the widespread adoption of HR crops is reducing its effectiveness in some situations. Glyphosate-resistant weeds have evolved where repeated applications of glyphosate have constituted the only weed-management tactic. Ten weed species in the United States have evolved resistance to glyphosate since the introduction of HR crops in 1996 compared with seven that have evolved resistance to glyphosate worldwide in areas not growing GE crops since the herbicide was commercialized in 1974. Furthermore, communities of weeds less susceptible to glyphosate are

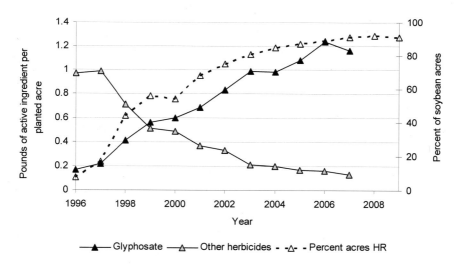

FIGURE S-1 Application of herbicide to soybean and percentage of acres of herbicide-resistant soybean.

NOTE: The strong correlation between the rising percentage of HR soybean acres planted over time, the increased applications of glyphosate, and the decreased use of other herbicides suggests but does not confirm causation between these variables.

SOURCE: USDA-NASS, 2001, 2003, 2005, 2007, 2009a, 2009b; Fernandez-Cornejo et al., 2009.

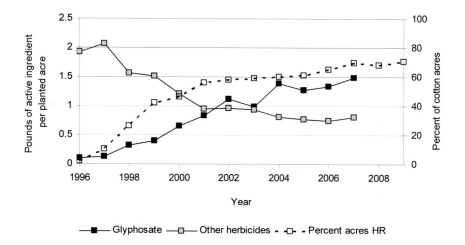

FIGURE S-2 Application of herbicide to cotton and percentage of acres of herbicide-resistant cotton.
NOTE: The strong correlation between the rising percentage of HR cotton acres planted over time, the increased applications of glyphosate, and the decreased use of other herbicides suggests but does not confirm causation between these variables.
SOURCE: USDA-NASS, 2001, 2003, 2005, 2007, 2009a, 2009b; Fernandez-Cornejo et al., 2009.

becoming established in fields planted with HR crops, particularly fields that are treated only with glyphosate.

Finding 2. The adoption of HR crops complements conservation tillage practices, which reduce the adverse effects of tillage on soil and water quality.

Farmers have traditionally used tillage to control weeds in their fields, interrupting weed life cycles before they can produce seeds for the following year. However, using tillage to help manage weeds reduces soil quality and increases soil loss from erosion. Tilled soil forms a crust, which reduces the ability of water to infiltrate the surface and leads to runoff that can pollute surface water with sediments and chemicals. Conservation tillage, which leaves at least 30 percent of the previous crop's residue on the field, improves soil quality and water infiltration and reduces erosion because more organic matter is left on the soil surface, thereby decreasing disruption of the soil. The adoption of HR crops

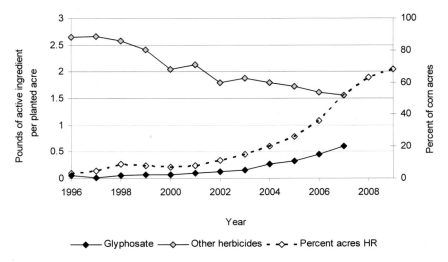

FIGURE S-3 Application of herbicide to corn and percentage of herbicide-resistant corn.
NOTE: The strong correlation between the rising percentage of HR corn acres planted over time, the increased applications of glyphosate, and the decreased use of other herbicides suggests but does not confirm causation between these variables.
SOURCE: USDA-NASS, 2001, 2003, 2005, 2007, 2009a, 2009b; Fernandez-Cornejo et al., 2009.

allows some farmers to substitute glyphosate application for some tillage operations as a weed-management tactic and thereby benefits soil quality and probably improves water quality, although definitive research on the latter is lacking. However, empirical evidence points to a two-way causal relationship between the adoption of HR crops and conservation tillage. Farmers who use conservation tillage are more likely to adopt HR crop varieties than those who use conventional tillage, and those who adopt HR crop varieties are more likely to practice conservation tillage than those who use non-GE seeds.

Finding 3. Targeting specific plant insect pests with Bt corn and cotton has been successful, and the ability to target specific plant pests in corn and cotton continues to expand. Insecticide use has decreased with the adoption of insect-resistant (IR) crops. The emergence of insect resistance to Bt crops has been low so far and of little economic or agronomic consequence; two pest species have evolved resistance to Bt crops in the United States.

Bt toxins, which are produced by the soil-dwelling bacterium *Bacillus thuringiensis*, are lethal to the larvae of particular species of moths, butter-flies, flies, and beetles and are effective only when an insect ingests the toxin. Therefore, crops engineered to produce Bt toxins that target specific pest taxa have had favorable environmental effects when replacing broad-spectrum insecticides that kill most insects (including beneficial insects, such as honey bees or natural enemies that prey on other insects), regard-less of their status as plant pests. The amounts of insecticides applied per planted acre of Bt corn and cotton have inverse relationships with the adoption of these crops over time (Figures S-4 and S-5), though a causative relationship has not been established or refuted because other factors influence pesticide-use patterns.

Since their introduction in 1996, the use of IR crops has increased rap-idly, and they continue to be effective. Data indicate that the abundance of refuges of non-Bt host plants and recessive inheritance of resistance are two key factors influencing the evolution of resistance. The refuge strate-gies mandated by the Environmental Protection Agency, and the promo-tion of such strategies by industry, likely contributed to increasing the use of refuges and to delaying the evolution of resistance to Bt in key pests.

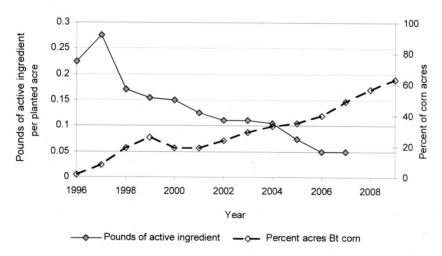

FIGURE S-4 Pounds of active ingredient of insecticide applied per planted acre and percent acres of Bt corn, respectively.
NOTE: The strong correlation between the rising percentage of Bt corn acres planted over time and the decrease in pounds of active ingredient per planted acre suggests but does not confirm causation between these variables.
SOURCE: USDA-NASS, 2001, 2003, 2005, 2007, 2009a, 2009b; Fernandez-Cornejo et al., 2009.

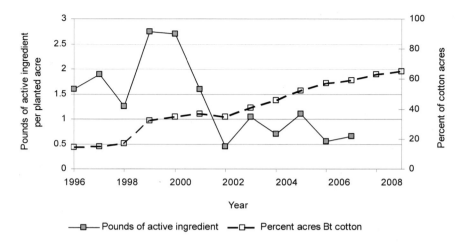

FIGURE S-5 Pounds of active ingredient of insecticide applied per planted acre and percent acres of Bt cotton, respectively.
NOTE: The strong correlation between the rising percentage of Bt cotton acres planted over time and the decrease in pounds of active ingredient per planted acre suggests but does not confirm causation between these variables.
SOURCE: USDA-NASS, 2001, 2003, 2005, 2007, 2009a, 2009b; Fernandez-Cornejo et al., 2009.

Nevertheless, some populations of two generalist pests have evolved resistance to Bt crops in the United States, although the agronomic and economic consequences appear to be minor. With the introduction of multiple Bt toxins in new hybrids or varieties, the probability of resistance to Bt crops is further reduced.

Finding 4. For the three major GE crops, gene flow to wild or weedy relatives has not been a concern to date because compatible relatives of corn and soybean do not exist in the United States and are only local for cotton. For other GE crops, the situation varies according to species. However, gene flow to non-GE crops has been a concern for farmers whose markets depend on an absence of GE traits in their products. The potential risks presented by gene flow may increase as GE traits are introduced into more crops.

Gene flow between many GE crops and wild or weedy relatives is low because GE crops do not have wild or weedy relatives in the United States or because the spatial overlap between a crop and its relatives is not extensive. How that relationship changes will depend on what GE crops are commercialized, whether related species with which they are capable of interbreeding are present, and the consequences of such interbreeding on weed management. Gene flow of approved GE traits into non-GE varieties of the same crops (known as adventitious presence) remains a serious concern for farmers whose market access depends on adhering to strict non-GE presence standards. Resolving this issue will require the establishment of thresholds for the presence of GE material in non-GE crops, including organic crops, that do not impose excessive costs on growers and the marketing system.

Economic Effects

The rapid adoption of GE crops since their commercialization indicates that the benefits to adopting farmers are substantial and generally outweigh additional technology fees for these seeds and other associated costs. The economic benefits and costs associated with GE crops extend beyond farmers who use the technology and will change with continuing adoption in the United States and abroad as new products emerge.

Finding 5. Farmers who have adopted GE crops have experienced lower costs of production and obtained higher yields in many cases because of more cost-effective weed control and reduced losses from insect pests. Many farmers have benefited economically from the adoption of Bt crops by using lower amounts of or less expensive insecticide applications, particularly where insect pest populations were high and difficult to treat before the advent of Bt crops.

The incomes of those who have adopted genetic-engineering technology have benefited from some combination of yield protection and lower costs of production. HR crops have not substantially increased yields, but their use has facilitated more cost-effective weed control, especially on farms where weeds resistant to glyphosate have not yet been identified. Lower yields were sometimes observed when HR crops were introduced, but the herbicide-resistant trait has since been incorporated into higher-yielding cultivars, and technological improvement in inserting the trait has also helped to eliminate the yield difference. In areas that suffer substantial damage from insects that are susceptible to the Bt toxins, IR crops have increased adopters' net incomes because of higher yields and reduced insecticide expenditures. Before the introduction of Bt crops, most farmers accepted yield losses to European corn borer rather than incur the expense and uncertainty of chemical control. Bt traits to address corn rootworm problems have lowered the use of soil-applied and seed-applied insecticides. In areas of high susceptible insect populations, Bt cotton has been found to protect yields with fewer applications of topical insecticides. More effective management of weeds and insects also means that farmers may not have to apply insecticides or till for weeds as often, and this translates into cost savings—lower expenditures for pesticides and less labor and fuel for equipment operations.

Finding 6. Adopters of GE crops experience increased worker safety and greater simplicity and flexibility in farm management, benefitting farmers even though the cost of GE seed is higher than non-GE seed. Newer varieties of GE crops with multiple GE traits appear to reduce production risk for adopters.

Farmers who purchase GE seed pay a technology fee—a means by which seed developers recover research and development costs and earn profits. GE seed is typically more expensive than conventional seed, and the net return in terms of higher yields and lower costs of production for a farmer considering adoption does not always offset the technology fee. However, studies have found that high rates of adoption of GE crops can be attributed in part to the value that farmers place on increased worker safety, perceived greater simplicity and flexibility in farm management (including more off-farm work opportunities), and lower production risk. Farmers and their employees not only face reduced exposure to the harsh chemicals found in some herbicides and insecticides used before the introduction of GE crops but have to spend less time in the field applying the pesticides. Because glyphosate can be applied over a fairly wide time-frame, farmers who use HR crops have greater flexibility regarding when they treat weeds in their fields. Those benefits must be balanced with the risk that such flexibility in application timing may reduce crop-yield potential attributable to weed interference. Newer GE varieties that have multiple pest-control traits may result in more consistent pest management and thus less yield variability, a characteristic that has substantial value for risk-averse producers. The value of those benefits may provide additional incentives for adoption that counteract the extra cost of GE seed.

Finding 7. The effect GE crops have had on prices received by farmers for soybean, corn, and cotton is not completely understood.

Studies suggest that the adoption of GE crops that confer productivity increases ultimately puts downward pressure on the market prices of the crops. However, early adopters benefit from higher yields or lower production costs more than nonadopters even with lower prices. The gains tend to dissipate as the number of adopters increases, holding technological progress constant. Thus, as the first adopters, U.S. farmers have generally benefited economically from the fact that GE crops were developed and commercialized in the United States before they were planted by farmers in other countries. The extent to which GE-crop adoption in developing countries will influence productivity and prices, and therefore U.S. farm incomes, is not completely understood. There is a paucity of studies of the economic effects of genetic-engineering technology in recent years even though adoption has increased globally.

Finding 8. To the extent that economic effects of GE-crop plantings on non-GE producers are understood, the results are mixed. By and large, these effects have not received adequate research.

Decisions made by adopters of GE crops can affect the input prices and options for both farmers who use feed and food products made with GE ingredients and farmers who have chosen not to grow GE seed or do not have the option available. The effects on those not using genetic-engineering technology have not been studied extensively. Livestock producers constitute a large percentage of corn and soybean buyers and therefore are major beneficiaries of any downward pressure on crop price due to the adoption of GE crops. Feed costs are nearly half the variable costs for livestock producers, so even moderate price fluctuations can affect their net incomes substantially. Livestock producers also benefit from increased feed safety due to reduced levels of mycotoxins in the grain. However, no quantitative estimation of savings to livestock operators due to the adoption of GE crops and the resulting effect on the profitability of livestock operations has been conducted. Similarly, a number of other economic effects predicted by economic theory have not been documented.

Favorable and unfavorable externalities are not limited to the cost and availability of inputs. To the extent that genetic-engineering technology successfully reduces pest pressure on a field and regionally, farmers of fields in the agricultural landscape planted with non-GE crops may benefit via lower pest-control costs associated with reductions in pest populations. However, nonadopters of genetic-engineering technology also could suffer from the development of weeds and insects that have acquired pesticide resistance in fields within the region planted to GE crops. When that happens, farmers might have to resort to managing the resistant pests with additional, potentially more toxic or more expensive forms of control, even though their practices may not have led to the evolution of resistance.

Inadvertent gene flow from GE to non-GE varieties of crops can increase production costs. Gene flow occurs through cross-pollination between GE and non-GE plants from different fields, co-mingling of GE seed with non-GE seed, and germination of seeds left behind (volunteers) after the production year. Similarly, if GE traits cross into weedy relatives, weed-control expenses will be higher for all fields on to which the weeds spread, whether a farmer grows GE crops or not. In addition, gene flow of GE traits into organic crops could jeopardize crop value by rendering outputs unsuitable for high-value foreign or other markets that limit or do not permit GE material in food products; the extent of that effect has not been documented during the last 5 years. On the other hand, the segregation of GE traits from organic production may have benefited organic producers by creating a market in which they can receive a premium for non-GE products.

Social Effects

The use of GE crops, like the adoption of other technologies at the farm level, is a dynamic process that both affects and is affected by the social networks that farmers have with each other, with other actors in the commodity chain, and with the broader community in which farm households reside. However, the social effects of GE-crop adoption have been largely overlooked.

Finding 9. Research on the dissemination of earlier technological development in agriculture suggests that favorable and unfavorable social impacts exist from the dissemination of genetic-engineering technology. However, these impacts have not been identified or analyzed.

Because GE crops have been widely adopted rapidly, it is reasonable to hypothesize that there have been social effects on adopters, nonadopters, and farmers who use GE products, such as livestock producers. For example, based on earlier research on the introduction of new technologies in agriculture, it is possible that certain categories of farmers (such as those with less access to credit, those with fewer social connections to university and private-sector researchers, or those who grow crops for smaller markets) might be less able to access or benefit from GE crops. The introduction of genetic-engineering technology in agriculture could also affect labor dynamics, farm structure, community viability, and farmers' relationships with each other and with information and input suppliers. However, the extent of the social effects of the dissemination of GE crops is unknown because little research has been conducted.

Finding 10. The proprietary terms under which private-sector firms supply GE seeds to the market has not adversely affected the economic welfare of farmers who adopt GE crops. Nevertheless, ongoing research

is needed to investigate how market structure may evolve and affect access to non-GE or single-trait seed. Furthermore, there has been little research on how increasing market concentration of seed suppliers affects overall yield benefits, crop genetic diversity, seed prices, and farmers' planting decisions and options.

During the 20th century, the U.S. seed industry evolved from small, family-owned businesses that multiplied seeds developed by university scientists to a market dominated by a handful of large, diversified companies. Universities still contribute to seed development, but seed companies have invested considerably in the research, development, and commercialization of patent-protected GE traits for large seed markets. Thus, corn, soybean, and cotton have received the bulk of private research attention in the last few decades. Large seed companies have not commercialized GE traits in many other crops because their market size has been insufficient to cover necessary research and development costs or because of concerns related to consumer acceptance and gene flow. Public research institutions continue to enhance the genetics of other crops, but full access to state-of-the-art technology (like genetic engineering) that may be beneficial to crops in smaller markets is often not available to public researchers because of patent protections.

Studies conducted in the first few years after the introduction of GE crops found no adverse effects on farmers' economic welfare from the consolidation of market power in the seed industry. However, the current developmental trajectory of GE-seed technology is causing some farmers to express concern that access to seeds without GE traits or to seeds that have only the specific GE traits that are of particular interest to farmers will become increasingly limited. Additional concerns are being raised about the lack of farmer input into and knowledge about which seed traits are being developed. Although the committee was not able to find published peer-reviewed material that documented the degree of U.S. farmers' access to non-GE seed and the quality of the seed, testimony provided to the committee suggests that access to non-GE or nonstacked seed may be restricted for some farmers or that available non-GE or nonstacked seed may be available in older cultivars that do not have the same yield characteristics as newer GE cultivars.

CONCLUSIONS AND RECOMMENDATIONS

Conclusion 1. Weed problems in fields of HR crops will become more common as weeds evolve resistance to glyphosate or weed communities less susceptible to glyphosate become established in areas treated exclusively with that herbicide. Though problems of evolved resistance

and weed shifts are not unique to HR crops, their occurrence, which is documented, diminishes the effectiveness of a weed-control practice that has minimal environmental impacts. Weed resistance to glyphosate may cause farmers to return to tillage as a weed-management tool and to the use of potentially more toxic herbicides.

A number of new genetically engineered HR cultivars are currently under development and may provide growers with other weed-management options when fully commercialized. However, the sustainability of those new GE cultivars will also be a function of how the traits are managed. If they are managed in the same fashion as the current genetically engineered HR cultivars, the same problems of evolved herbicide resistance and weed shifts may occur. Therefore, farmers of HR crops should incorporate more diverse management practices, such as herbicide rotation, herbicide application sequences, and tank-mixes of more than one herbicide; herbicides with different modes of action, methods of application, and persistence; cultural and mechanical control practices; and equipment-cleaning and harvesting practices that minimize the dispersal of HR weeds.

Recommendation 1. Federal and state government agencies, private-sector technology developers, universities, farmer organizations, and other relevant stakeholders should collaborate to document emerging weed-resistance problems and to develop cost-effective resistance-management programs and practices that preserve effective weed control in HR crops.

Conclusion 2. Given that agriculture is the largest source of surface water pollution, improvements in water quality resulting from the complementary nature of herbicide-resistance technology and conservation tillage may represent the largest single environmental benefit of GE crops. However, the infrastructure to track and analyze these effects is not in place.

Recommendation 2. The U.S. Geological Survey and companion federal and state environmental agencies should receive the financial resources necessary to document the water quality effects related to the adoption of GE crops.

Conclusion 3. The environmental, economic, and social effects on adopters and nonadopters of GE crops has changed over time, particularly because of changes in pest responses to GE crops, the consolidation of the seed industry, and the incorporation of GE traits into most varieties of corn, soybean, and cotton. However, empirical research into the environmental and economic effects of changing market conditions and farmer practices have not kept pace. Furthermore, little work has been conducted regarding the effects on livestock producers and non-adopters and on the social impacts of GE crops. Issues in need of further investigation include the costs and benefits of shifts in pest management for non-GE producers due to the adoption of GE crops, the value of market opportunities afforded to organic farmers by defining their products as non-GE, the economic impacts of GE-crop adoption on livestock producers, and the costs to farmers, marketers, and processors of the presence of approved or unapproved GE traits and crops in products intended for restricted markets. As more GE traits are developed and inserted into existing GE crops or into other crops, understanding the impacts on all farmers will become even more important to ensuring that genetic-engineering technology is used in a way that facilitates environment, economic, and social sustainability in U.S. agriculture.

Recommendation 3. Public and private research institutions should allocate sufficient resources to monitor and assess the substantial environmental, economic, and social effects of current and emerging agricultural biotechnology on U.S. farms so that technology developers, policy makers, and farmers can make decisions that ensure genetic engineering is a technology that contributes to sustainable agriculture.

Conclusion 4. Commercialized GE traits are targeted at pest control, and when used properly, they have been effective at reducing pest problems with economic and environmental benefits to farmers. However, genetic engineering could be used in more crops, in novel ways beyond herbicide and insect resistance, and for a greater diversity of purposes. With proper management, genetic-engineering technology could help address food insecurity by reducing yield losses through its introduction into other crops and with the development of other yield protection traits like drought tolerance. Crop biotechnology could also address "public goods" issues that will be undersupplied by the market acting alone. Some firms are working on GE traits that address public goods issues. However, industry has insufficient incentive to invest enough in research and development for those purposes when firms cannot collect revenue from innovations that generate net benefits beyond the farm. Therefore, the development of these traits will require greater collaboration between the public and private sectors because the benefits extend beyond farmers to the society in general. The implementation of a targeted and tailored regulatory approach to GE-trait development and commercialization that meets human and environmental safety standards while minimizing unnecessary expenses will aid this agenda (Ervin and Welsh, 2006).

Recommendation 4. Public and private research institutions should be eligible for government support to develop GE crops that can deliver valuable public goods but have insufficient market potential to justify private investment. Intellectual property patented in the course of developing major crops should continue to be made available for such public goods purposes to the extent possible. Furthermore, support should be focused on expanding the purview of genetic-engineering technology in both the private and public sectors to address public goods issues. Examples of GE-crop developments that could deliver such public goods include but are not limited to

- *plants that reduce pollution of off-farm waterways through improved use of nitrogen and phosphorus fertilizers,*
- *plants that fix their own nitrogen and reduce pollution caused by fertilizer application,*
- *plants that improve feedstocks for renewable energy,*
- *plants with reduced water requirements that slow the depletion of regional water resources,*
- *plants with improved nutritional quality that deliver health benefits, and*
- *plants resilient to changing climate conditions.*

REFERENCES

Ervin, D., and R. Welsh. 2006. Environmental effects of genetically modified crops: Differentiated risk assessment and management. In *Regulating agricultural biotechnology: Economics and policy.* eds. R.E. Just, J.M. Alston, and D. Zilberman, pp. 301–326. New York: Springer.

Fernandez-Cornejo, J., R. Nehring, E.N. Sinha, A. Grube, and A. Vialou. 2009. Assessing recent trends in pesticide use in U.S. agriculture. Paper presented at the 2009 Annual Meeting of the Agricultural and Applied Economics Association (Milwaukee, WI, July 26–28, 2009). Available online at http://aeconsearch.umn.edu/handle/49271. Accessed June 16, 2009.

USDA-NASS (U.S. Department of Agriculture–National Agricultural Statistics Service). 2001. Acreage. June 29. Cr Pr 2-5 (6-01). Washington, DC. Available online at http://usda.mannlib.cornell.edu/usda/nass/Acre//2000s/2001/Acre-06-29-2001.pdf. Accessed April 14, 2009.

———. 2003. Acreage. June 30. Cr Pr 2-5 (6-03). Washington, DC. Available online at http://usda.mannlib.cornell.edu/usda/nass/Acre//2000s/2003/Acre-06-30-2003.pdf. Accessed April 14, 2009.

———. 2005. Acreage. June 30. Cr Pr 2-5 (6-05). Washington, DC. Available online at http://usda.mannlib.cornell.edu/usda/nass/Acre//2000s/2005/Acre-06-30-2005.pdf. Accessed April 14, 2009.

———. 2007. Acreage. June 29. Cr Pr 2-5 (6-07). Washington, DC. Available online at http://usda.mannlib.cornell.edu/usda/nass/Acre//2000s/2007/Acre-06-29-2007.pdf. Accessed April 14, 2009.

————. 2009a. Data and statistics: Quick stats. Washington, DC. Available online at http://www.nass.usda.gov/Data_and_Statistics/Quick_Stats/index.asp. Accessed June 22, 2009.

————. 2009b. Acreage. June 30. Cr Pr 2-5 (6-09). Washington, DC. Available online at http://usda.mannlib.cornell.edu/usda/current/Acre/Acre-06-30-2009.pdf. Accessed November 24, 2009.

1

Introduction

Historians often link the advent of human civilizations with the transition of human societies from food collection primarily through hunting and gathering to food production in established agricultural systems. In a pattern of parallel development, early agricultural systems began emerging in separate regions during the Neolithic period some 10,000 years ago (Mazoyer and Roudart, 2006). Crop-improvement practices based on identification and selection of the best plant varieties appear to date back to the early days of agriculture itself. Similarly, early pastoralists engaged in selective animal breeding. That those practices were recognized as important in the development of ancient human civilizations is apparent in the preservation of instructions on plant breeding in writing, such as in the works of Virgil and Theopastus (Vavilov, 1951). In the broadest sense, the term *biotechnology* can encompass a wide array of procedures used to modify organisms according to human needs. It can be argued that early agriculturalists engaged in a simple form of biotechnology (Kloppenburg, 2004) in developing the intention and the techniques to improve plant varieties and animal species.

Although the process of plant and animal improvement has been continuous throughout the history of agriculture, some historical periods can be identified as singularly transformative. For example, a major agricultural revolution took place in Europe from the 16th to the 19th centuries. It was characterized in part by the extensive use of plants and animals that had been imported from the Americas (Crosby, 2003) and by animal-drawn cultivation and the use of fertilizers, the latter permitting cereal and feed-grain cultivation without fallowing (Mazoyer and Roudart, 2006). That revolution led to important increases in the food supply and thus ultimately permitted increased population growth.

Another important change in agriculture resulted from the application of an increasingly scientific approach to plant breeding, which developed from the recognition of the cell as the primary unit of all living organisms in the 1830s (Vasil, 2008) and the work of Mendel (Kloppenburg, 2004). With the rediscovery of Mendel's principles of genetics in the early 1900s, progress in plant and animal breeding was accelerated. The continuous growth in crop yields and agricultural productivity during the 20th century owes much to those biological discoveries and to a series of mechanical and chemical innovations driven by agricultural research and development.

One of the more significant innovations in plant breeding during the 20th century was the development of hybrid crops, particularly corn, in the United States. Hybrid corn varieties, which are developed from crossing different inbred lines, out-yield pure inbred lines, though the seeds produced by hybrid varieties yield poorly. When corn hybrids were first developed, they had no discernible yield advantage over the existing open-pollinated corn varieties of the time (Lewontin, 1990). However, seed companies were motivated to develop high-yielding hybrid varieties; saving and planting the seeds of hybrid corn did not produce equal yields, so seed companies had a financial incentive to invest in these varieties. The research and development efforts devoted to hybrid corn produced tremendous yield improvements over the last 70 years. It is unclear if the same amount of investment could have resulted in similar yield increases for open-pollinated varieties; regardless, because of their limited potential for return on financial investment, efforts to develop high-yielding open-pollinated varieties were not made. Modern hybrids, which have been bred to allocate more of their energy to producing grain rather than stover (leaves and stalks), also demonstrate an ability to maintain high grain production in densely planted fields (Liu and Tollenaar, 2009), and they can exhibit increased tolerance to environmental stresses (such as drought, cold, and light availability).

Plant breeders in the 20th century also identified varieties of wheat and rice with shorter stalks and larger seed heads. They were crossed with relatives to create semidwarf wheat and rice varieties, which produced greater yields in part because they responded well to applications of nitrogen and did not lodge despite having heavier seed heads. The development of semidwarf wheat and rice spurred the Green Revolution of the 1960s and 1970s in developing countries (Conway, 1998). Such improvements in plant breeding increased global crop yields in rice and wheat substantially in countries with suitable growing conditions and markets.

Recent developments in scientific plant breeding have resulted from discoveries in molecular and cellular biology in the second half of the 20th century that laid the foundation for the development of genetically engineered plants. In 1973, the American biochemists Stanley Cohen and Herbert Boyer were among the first scientists to transfer a gene between unrelated organisms successfully. They cut DNA from an organism into fragments, rejoined a subset of those fragments, and added the rejoined subset to bacteria to reproduce. The replicated DNA fragments were then spliced into the genome of a cell from a different species, and this created a transgenic organism, that is, an organism with genes from more than one species. Before the advent of genetic engineering, plant tissue-culture technology expanded the array of available genetic material beyond what was possible with traditional plant breeding by manipulating the fertilization and embryos of crosses between more distantly related species (Brown and Thorpe, 1995). DNA-recombination techniques opened the possibility of augmenting plant genomes with desirable traits from other species and thus took the science of plant breeding to a stage in which improvement is constrained not by the limits of genetic traits within a particular species but rather by the limits of discovery of genes and their transfer from one species to another to confer desired characteristics on a particular crop.

COMMITTEE CHARGE AND APPROACH

The committee's study was the first comprehensive assessment of the impacts of the use of genetically engineered (GE) crops on farm sustainability in the United States. The most up-to-date, available scientific evidence from all regions was used to assemble a national picture that would reflect important variations among regions. Box 1-1 presents the formal statement of task assigned to the committee.

In conducting its task, the committee interpreted the term *sustainability* to apply to the environmental, economic, and social impacts of genetic-engineering technology at the farm level. That interpretation is in line with the federal government's definition of *sustainable agriculture*, which is "an integrated system of plant and animal production practices having a site-specific application that will over the long term:

1. Satisfy human food and fiber needs.
2. Enhance environmental quality and the natural resource base upon which the agriculture economy depends.
3. Make the most efficient use of nonrenewable resources and on-farm resources and integrate, where appropriate, natural biological cycles and controls.

BOX 1-1
Statement of Task

An NRC committee will study the farm-level impacts of biotechnology, including the economics of adopting genetically engineered crops, changes in producer decision making and agronomic practices, and farm sustainability.
The study will:

• review and analyze the published literature on the impact of GE crops on the productivity and economics of farms in the United States;
• examine evidence for changes in agronomic practices and inputs, such as pesticide and herbicide use and soil and water management regimes;
• evaluate producer decision making with regard to the adoption of GE crops.

In a consensus report, the committee will present the findings of its study and identify future applications of plant and animal biotechnology that are likely to affect agricultural producers' decision making in the future.

4. Sustain the economic viability of farm operations.
5. Enhance the quality of life for farmers and society as a whole."
(Food, Agriculture, Conservation, and Trade Act of 1990)

This definition conceives of sustainable farming systems that address salient environmental, economic, and social aspects and their interrelationships.

The report explores how GE crops contribute to achieving several of the conditions enumerated above. Farmers must continually adapt in response to environmental, economic, and social conditions by learning and adopting new practices. Adopting GE crops is one option some farmers make in adapting to changing conditions.

Though the three aspects of sustainability often interact with one another, the report organizes each in a separate chapter to facilitate access to the information. The chapter on production economics follows the environmental chapter because many of the economic gains and losses that farmers experience with GE crops result from changes occurring within the farm environment from GE-crop adoption. The chapter on social effects is brief because of a lack of published literature on the subject. Nevertheless, the committee deemed this aspect important to include for two reasons. First, social impacts are widely considered to be a necessary element in the definition of sustainability as noted earlier. Second, with the sizable shift in cropping practices and systems to genetic-engineering technology (and the prospect of more GE crops to come), the marked expansion of private-sector control of intellectual property related to seeds, and a growing concentration of private-sector seed companies, it is the committee's estimation that GE crops have had and will continue to have social repercussions at the farm and community levels. The committee agreed that the report should draw attention to the need for research in this area. In this vein, the report highlights issues on which insufficient information is available for drawing firm conclusions. The final chapter summarizes the main findings of the assessment and discusses the potential for future GE crops to address emergent food, energy, and environmental challenges.

The committee interpreted the statement of task to be retrospective in nature, examining the sustainability effects of GE crops on U.S. farms since their commercialization. For that reason the committee focused in large part on the experiences of soybean, corn, and cotton producers because GE varieties of those crops have been widely adopted by farmers, those crops are planted on almost half of U.S. cropland, and most research on genetic-engineering technology in agriculture has targeted those three crops. However, the committee recognized that most farmers have been affected by the widespread adoption of GE crops, even if they have chosen not to adopt them or have not had the option to adopt them. The report examined the effects of genetic-engineering technology on those producers as well. Because the study was retrospective and focused on the experience of U.S. farmers, the adoption of GE crops in other countries entered into the analysis only if U.S. farmers have experienced effects of such adoption, and the committee restricted its speculations on the future applications and implications of genetic-engineering technology to the final chapter.

The National Research Council supported the study to expand its contributions to the understanding of agricultural biotechnology. Committee members were chosen because of their academic research and experience on the topic. Experts were selected from the fields of weed science, agricultural economics, ecology, rural sociology, environmental economics, entomology, and crop science. To prepare its report, the committee reviewed previous studies and scientific literature on farmers' adoption of genetic-engineering technology, the impacts of such technology on non-GE farmers, and environmental impacts of GE crops. It also examined historical and current statistical data on the adoption of GE crops in the United States. The committee acknowledges that GE crops in

U.S. agriculture continue to stir controversy around scientific issues and ideological viewpoints. With this in mind, the committee kept its focus on scientific questions and adopted an evidentiary standard of using peer-reviewed literature upon which to base its conclusions and recommendations. It refrained from analyzing ideological positions, either in support of or against the technology, in order to remain as impartial as possible.

STUDY FRAMEWORK

An analysis of the farm-level sustainability impacts of GE crops requires a framework that integrates all salient factors that motivate their use. We use the principal theories applied to agricultural technology adoption to construct a framework that identifies the qualitative factors that affect U.S. farmers' decisions to use genetic-engineering technology. With an understanding of the adoption and use processes, we then outline an evaluation framework that spans environmental, economic, and social dimensions as noted above.

Two main theories help in building a framework for analyzing a farmer's decision to adopt a particular GE crop. First, "diffusion" theory seeks to explain people's propensities to adopt innovations as communicated through particular channels and within particular social systems (Rogers, 2003). Second, "threshold" theory delves deeper into the economic influences on farmer decisions by considering the heterogeneity in farm sizes, in agronomic conditions (climate, soil, water availability, and pest pressure), in forms of human capital that influence learning by doing and using, and in operator values (Feder et al., 1985; Foster and Rosenzweig, 1995; Fischer et al., 1996; Marra et al., 2001; Sunding and Zilberman, 2001). Incorporating those factors allows a better qualitative understanding of the dynamics of the spread of the technologies across the landscape and of their impacts. Together, the diffusion and threshold theories point to five sets of factors that exert influences on a farmer's decision to use genetic-engineering technology:

1. Productivity (yield) effects.
2. Market structure and price effects.
3. Production-input effects.
4. Human capital and personal values.
5. Information and social networks.

Productivity Effects

Genetic-engineering technology can directly and indirectly affect crop yields, either positively or negatively, as explained in Chapter 3 in more detail. The direct route stems from the effect on a cultivar after the insertion of one or more traits through genetic engineering. The indirect effect is related to the ability of a GE crop to decrease pest damage (Lichtenberg and Zilberman, 1986a). Just as natural-resource conditions, including pest pressures, vary among fields, farms, and regions, so will the indirect effects on yield and the rate of adoption of GE crops. The technologies tend to be adopted in locations whose agrophysical conditions—such as land quality, climate, and vulnerability to pests—lead to productivity gains (Marra et al., 2003; Zilberman et al., 2003). In addition to effects on quantity, genetic engineering may affect the quality of a crop, which influences its value.

Market-Structure and Price Effects

Farmers who are deciding whether to grow GE crops must consider their access to domestic and foreign markets. Differential access may stem from country regulations on the entry of GE crops into their markets or from lack of market infrastructure (for example, segmentation of GE and non-GE product chains). Farmers who choose to grow GE crops may experience higher or lower prices than if they grow non-GE crops. For example, if enough farmers adopt a GE crop and yields increase substantially because of direct or indirect effects, crop prices may be forced down by increased supplies, other characteristics remaining the same. Consumers of GE crops may benefit from the lower prices, though some consumers may be willing to pay more for non-GE crops for personal reasons, and this may create a premium for non-GE crops. Under other circumstances, global demand increases may absorb most or all of the increase in supply, in which case prices would not decline (see Chapter 3).

Market access and price effects alter farmers' revenues and profitability and thus their disposition to adopt GE crops. The organizational hierarchy of the commodity chain and the nature of farm policies can create structural conditions that act as impediments to or inducers of adoption of a technology (Mouzelis, 1976; Bonanno, 1991; Friedland, 2002; Kloppenburg, 2004). For example, the development of crops with more than one GE trait may create a structural condition for some farmers whereby they may have to pay for traits that they do not need in order to gain access to the traits that they desire (see Chapter 4).

Production-Input Effects

The adoption and use of GE crops can precipitate changes in the types, amounts, and timing of pesticide use and in the types, frequency, and timing of tillage operations; both can affect machinery require-ments. Those changes are referred to as substitution effects; an example is the replacement of some pesticides with a GE crop (Lichtenberg and Zilberman, 1986b). A shift in labor requirements is another potentially important production-input effect (Fernandez-Cornejo and Just, 2007). The availability and quality of GE and non-GE seeds may affect a farmer's decision to use either. For example, the commercial success of the applica-tion of GE soybean and corn in the 1990s was accompanied by increased consolidation and vertical integration in the seed industry (Fernandez-Cornejo, 2004). Indeed, by 1997, two firms captured 56 percent of the U.S. corn-seed market, and this share has increased even more in recent years (see "Interaction of the Structure of the Seed Industry and Farmer Decisions" in Chapter 4) (Boyd, 2003). The changes in genetic-engineering technology and seed-industry structure may help to explain anecdotal statements about the reduced availability of some non-GE seed varieties in recent years (Hill, personal communication). However, the committee is not aware of any published research confirming the link between seed-industry structure and seed availability.

Human Capital and Personal Values

Every major study of agricultural-technology adoption has found that at least some aspects of human capital play a role in the process. Fre-quently, the more education or experience a farmer has, the more likely he or she is to adopt a new technology. Educational achievement and years of experience in farming are thought to be proxies for a potential adopter's ability to learn quickly how to adapt the new technology to the farm oper-ation and to use it to its greatest advantage. As noted above, the process of learning and adaptation is critical to the development of more sustainable farming systems. Farmers also may hold personal values that affect their decisions to use GE crops beyond the financial effects that may flow from productivity, value, and production input. A person's values define prefer-ences and have been shown to influence decisions on genetic-engineering development and applications (Piggott and Marra, 2008; Buccola et al., 2009). Examples of personal values include aversion to general and spe-cific risks, preference for environmental stewardship, and ideological positions about agricultural systems. An example of the influence of risk aversion is some farmers' preference for GE crops if they reduce the vari-ability of yields because they improve control of pests. Such risk reduction can motivate adoption of GE varieties by risk-averse farmers and may also lead to an increase in use of complementary practices, such as no-till planting (Alston et al., 2002; Piggott and Marra, 2007).

Information and Social Networks

Decisions of whether to adopt GE crops hinge on the quantity and quality of farmers' information about the characteristics and performance of the technologies. Information from formal sources, such as the agricultural media, on GE traits' technical aspects, economic implications, and prospects can shape farmers' views. Informal sources probably also speed or slow the adoption of GE crops (Wolf et al., 2001; Just et al., 2002). Social networks can have favorable or unfavorable effects not only on the adoption of technologies but also on the sharing of knowledge about GE and non-GE crops and on the development of new technologies and management strategies (Arce and Marsden, 1993; Busch and Juska, 1997; Hubbell et al., 2000). They can also mitigate potentially negative social impacts of GE-crop adoption. Recognition of the importance of social networks has been enhanced by studies of the processes associated with the use of alternative agricultural practices (Storstad and Bjørkhaug, 2003; Morgan et al., 2006). Insights derived from the study of social networks also may have great relevance to the development and dispersion of genetic-engineering technology.

Figure 1-1 portrays the influences of the different factors on GE-crop adoption decisions and the resultant impacts on environmental, economic, and social conditions. This conceptual model shows that factors under the control of the farmer, such as human capital, and outside their control, such as market prices, come together to influence the GE-crop adoption decision process, depicted by the central box in the figure. It also shows how the factors, up to this point presented as having distinct effects, may influence each other. Examples of potential interactions include the effects of information and social networks on personal values and production inputs and the effect of production-input substitution on productivity. Other impacts of decisions related to GE crops (for example, the environmental effect of pest population changes) may feed back to some influencing factors, such as production inputs. As discussed later in this chapter, empirical studies have found that factors in each of the categories have influenced GE-crop adoption patterns. However, it is not possible to rank the magnitude of influences in a general sense. Rather, we expect that the different factors will vary in influence across types of farms, geographic regions, and specific crop applications. For example, if a certain pest infestation is severe in a region, then the productivity gains from adopting a GE crop may far outweigh the influence of personal values of the adopter. In another case where pest pressures are moderate compared to those in other regions, functioning information and social networks may influence the speed and rate of adoption of genetic-engineering technology.

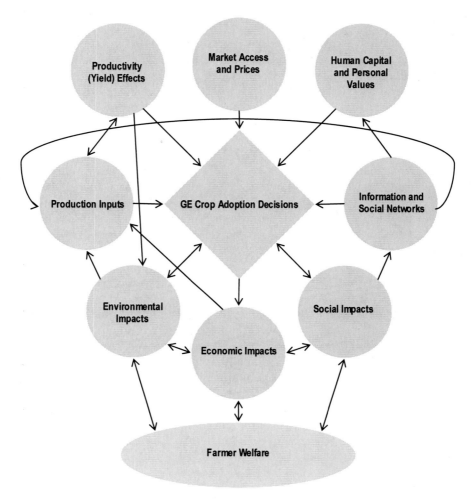

FIGURE 1-1 Genetically engineered crop adoption and impact framework.

GENETICALLY ENGINEERED TRAITS IN CROPS

For agricultural crops, the first generation of genetic engineering has targeted traits that increase the efficacy of pest control. Since the introduction of GE crops, new seeds have provided pest control in one or more of three forms:

- Herbicide resistance.
- Insect resistance.
- Virus resistance.

The terms *resistance* and *tolerance* are often used interchangeably in the literature. *Tolerance* implies that a crop is affected by a pesticide but has a means to naturally survive the potential damage sustained. This report uses the more precise term *resistance* because altered genes either allow a plant to generate its own insecticide or prevent herbicides from damaging the plant (Roy, 2004).

GE herbicide-resistant (HR) crops contain transgenes that enable survival of exposure to particular herbicides. In the United States, crops are available with GE resistance to glufosinate and glyphosate, but most HR crops grown in the United States are resistant only to glyphosate, a nonselective chemical that has a low impact on the environment. Glyphosate inhibits the enzyme 5-enolpyruvyl-shikimate-3-phosphate synthase (EPSPS), which is part of the shikimate pathway in plants. The shikimate pathway helps produce aromatic amino acids; it is speculated that glyphosate kills a plant either by reducing aromatic amino acid production and adversely affecting protein synthesis or by increasing carbon flow to the glyphosate-inhibited shikimate pathway, causing carbon shortages in other pathways (Duke and Powles, 2008). The susceptibility of EPSPS to the chemical and the relative ease with which it is taken up by a plant make glyphosate an extremely effective herbicide. It presents a low threat of toxicity to animals in general because they do not have a shikimate pathway for protein synthesis (Cerdeira and Duke, 2006). Glyphosate also has low soil and water contamination potential because it binds readily to soil particles and has a relatively short half-life in soil (Duke and Powles, 2008).

Insect-resistant (IR) plants grown in the United States have genetic material from the soil-dwelling bacterium *Bacillus thuringiensis* (Bt) incorporated into their genome that provides protection against particular insects. Bt produces a family of endotoxins, some of which are lethal to particular species of moths, flies, and beetles. An insect's digestive tract activates the ingested toxin, which binds to receptors in the midgut; this leads to the formation of pores, cell lysis, and death. Individual Bt toxins have a narrow taxonomic range of action because their binding to midgut receptors is specific; the toxicity of Bt crops to vertebrates and many nontarget arthropods and other invertebrates in U.S. agricultural ecosystems is effectively absent. The first Bt crops that were introduced produced only one kind of Bt toxin. More recent varieties produce two or more Bt toxins; this enhances control of some key pests, allows control of a wider array of insects, and can contribute to delaying the evolution of resistance in target pests while reducing refuge size.

Gene sequences of pathogenic viruses have been inserted into crops to confer protection against related viruses—to make them virus-resistant (VR). Most transgenic VR plants resist viruses through gene silencing, which occurs when transcription of a transgene induces degradation of the genome of an invading virus. Potential unwanted environmental effects of VR crops include exchanges between viral pathogens and transgene products that could increase the virulence of viral pathogens, food allergenicity, and transgene movement through pollen, which can create VR weeds. Adverse environmental effects of commercialized VR plants have not been found (Fuchs and Gonsalves, 2008).

HR and IR crops, having been the principal targets of most efforts to develop GE crop varieties, account for the bulk of acres planted in GE crops in the United States. Consequently, this report focuses on farmers' experiences with these types of GE crops. HR varieties of soybean, corn, cotton, canola, and sugar beets and IR varieties of corn and cotton were grown commercially in 2009. Herbicide resistance and insect resistance are not mutually exclusive; a number of crop varieties that contain both types of resistance have been developed. GE corn and cotton may also express more than one type of Bt trait. Seeds with multiple GE characteristics are referred to as "stacked cultivars."

Herbicide resistance and insect resistance were commercialized because of the relative simplicity in gene transfer and the utility for farmers. The expression of those traits requires manipulation of the genetic code at only one site, a relatively straightforward process compared with such traits as drought tolerance, which involve the action of many genes. Furthermore, because corn, soybean, and cotton production accounts for the bulk of pesticide expenditures in the United States (Figure 1-2), herbicide resistance and insect resistance provided important market opportunities. Those GE crops fit easily into the traditional pest-management approach of mainstream U.S. agriculture: reliance on the continual emergence of technological advances to address pest problems, particularly after development of resistance to an earlier innovation. Therefore, the familiarity of the chemicals involved, the size of the market for the seeds of and pesticides for GE crops, and the ease of manipulation of the genes for the traits contributed to HR and IR seeds' being the first GE products to emerge in large-scale agriculture.

ADOPTION AND DISTRIBUTION OF
GENETICALLY ENGINEERED CROPS

Crops with GE traits aimed primarily at pest control have been widely adopted in the United States by farmers of corn, cotton, soybean, canola, and sugar beet and have caused substantial changes in farm-management practices and inputs, such as changes in pesticide use. In 2009, almost half of U.S. cropland was planted with GE seed, even though the technology had been available to farmers only since the mid-1990s and only a few

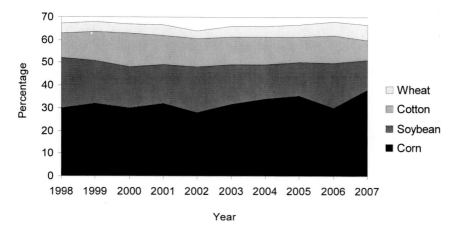

FIGURE 1-2 Share of major crops in total pesticide expenditures, 1998–2007.
NOTE: Includes expenditures in herbicides, insecticides, and fungicides. Genetically engineered trait technology fees are not included.
SOURCE: Fernandez-Cornejo et al., 2009.

crops have experienced commercial success (USDA-NASS, 2009b). U.S. farmers planted 158 million acres of GE crops in 2009—nearly half of all the GE-crop acres in the world (James, 2009). Rates of adoption have been influenced by the type of crop, the trait expressed in the crop, and the pest pressures occurring on the farm. For example, adoption of cultivars with Bt traits has been most rapid and widespread in areas prone to insect infestations that can be curbed by the endotoxins present in GE crops.

The committee chose to concentrate its study on the farm-level effects of GE soybean, corn, and cotton because these crops are grown on nearly half of U.S. cropland (USDA-NASS, 2009b) and because over 80 percent of these crops are genetically engineered (Figure 1-3). The high level of adoption and the large-scale planting of those crops mean they have a substantially greater cumulative impact on farm-level sustainability compared to other GE crops, which may be widely adopted but are planted on few acres or may be adopted by only a small percentage of growers. Additionally, there are GE crops that have been commercialized but were not sold in 2009 for business or legal reasons. Those crops are discussed in the report, but they are not its primary focus (Box 1-2).

Soybean

Soybean resistant to the herbicide glyphosate was first introduced in the United States in 1996. Just 4 years later, the GE cultivars accounted for

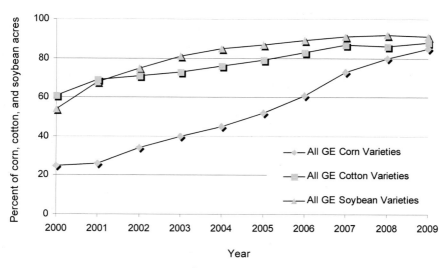

FIGURE 1-3 Nationwide acreage of genetically engineered soybean, corn, and cotton as a percentage of all acreage of these crops.
SOURCE: USDA-NASS, 2001, 2003, 2005, 2007, 2009b.

54 percent of all soybean acres planted (Table 1-1). A major factor in the rapid adoption was the superior control of a broad spectrum of weeds, including many problematic weeds, with a single timely application of glyphosate, especially in northern latitudes (Corrigan and Harvey, 2000; Mulugeta and Boerboom, 2000; Wiesbrook et al., 2001; Bradley et al., 2007). Other factors contributing to the rapid early adoption were the perceived simplicity and the relative safety of use (one application with a single herbicide compared with tank-mixed herbicides applied twice or more), lack of crop injury, lack of residual soil activity and potential injury to a succeeding crop, and the relatively low cost of glyphosate (Scursoni et al., 2006). Weeds resistant to glyphosate evolved in other instances, which required additional herbicides to be applied with glyphosate—roughly 25 percent of the acreage in some regions (Dill et al., 2008)—probably in an effort to prevent the evolution of such weeds. Despite that development in some regions, growers continued to adopt glyphosate-resistant varieties, as indicated by the fact that 91 percent of the acreage was planted to HR varieties in 2009 (Table 1-1). Adoption of HR soybean has been widespread in all regions since the early-adoption phase, and almost all soybean-producing states now hover around 90 percent adoption (Table 1-1, Figure 1-4).

BOX 1-2
Other Commercialized Genetically Engineered Crops

GE varieties of crops other than corn, soybean, and cotton have been developed and produced in the United States. However, some are planted on relatively small acreage and often in specific locations (for example, papaya in Hawaii), and others are no longer available commercially (for example, GE potato). These are highlighted below, but they are not the focus of the report because those in production account for only a small percentage of U.S. agriculture in terms of land and revenue, and those not commercially available were only sold for a short time. The latter demonstrates commercial viability can depend on the willingness of farmers to adopt the product and the willingness of processors and consumers to accept it.

Minor Crops Widely Planted in Genetically Engineered Varieties by U.S. Farmers

Canola. Canola is widely grown in Canada but is a minor crop in the United States. HR canola was commercialized in the mid-1990s. Since 2005, HR canola has accounted for almost half the acres planted in North Dakota, which makes up more than 87 percent of U.S. canola (USDA-NASS, 2009b). GE glyphosate-resistant and glufosinate-resistant canola cultivars accounted for 65 percent and 32 percent of the canola acres planted in the United States in 2006 (Howatt, personal communication).

Sugar beet. HR sugar beet was approved for commercial use in 1998, but concerns about marketplace acceptance precluded the commercial release of the transgenic cultivars (Duke, 2005). Transgenic glyphosate-resistant sugar beet was commercially grown in 2009 and reportedly was widely adopted (Stachler, personal communication). Economic studies suggested that the transgenic cultivars would improve profitability compared with conventional cultivars (Kniss et al., 2004; Gianessi, 2005). The potential for hybridization between HR sugar beet and weedy beet that occurs in the fields and nearby has led to concerns about future weed problems in Europe (where GE sugar beet is not commercialized), but hybridization is a minor problem in the United States (Stewart et al., 2003; Andersen et al., 2005). Indeed, spatial overlap between sugar beet production and weedy beet is limited to a few California counties (Calflora, 2009).

Papaya. Papaya is a small tree grown throughout the tropics for its fruit. Researchers at Cornell University and the University of Hawaii developed VR papaya by transforming the coat protein gene of the

continued

BOX 1-2 Continued

papaya ringspot virus to confer resistance against this devastating pathogen (Fuchs and Gonsalves, 2008). Transgenic papaya was first planted in 1998 in Hawaii, where it has contributed to sustaining the culture of this fruit. About 80 percent of papaya production in Hawaii is from transgenic plants (Fuchs and Gonsalves, 2008; Stokstad, 2008).

Minor Crops with Genetically Engineered Varieties Not Widely Adopted by U.S. Farmers

Squash. Transgenic VR squash lines were transformed with the coat protein genes of the watermelon mosaic virus, zucchini yellow mosaic virus, and cucumber mosaic virus (Fuchs and Gonsalves, 2008). The ZW-20 squash line is resistant to the watermelon and zucchini yellow mosaic viruses, and CZW-3 resists the watermelon, zucchini yellow, and cucumber mosaic viruses. ZW-20 and CZW-3 were deregulated in 1994 and 1996, respectively, and commercially grown soon thereafter. Those lines were also crossed to other squash cultivars to produce new lines of VR squash. Transgenic squash accounted for about 12 percent of total U.S. squash production in 2005. Most acreage of transgenic squash is in New Jersey, Florida, Georgia, South Carolina, and Tennessee (Fuchs and Gonsalves, 2008).

Sweet corn. Sweet corn was planted on 617,350 acres in 2008 (USDA-NASS, 2009a). A little less than half, or 246,600 acres, were planted for the fresh market; sweet corn from the remaining acres was for processing. Of the fresh market acres, about 20,000 were planted to GE varieties with high Bt protection against corn earworm and European corn borer and moderate protection against fall armyworm (see Table 1-2; Lynch et al., 1999; Mason, personal communication). Bt sweet corn varieties are typically marketed directly to the consumer; processors have been reluctant to purchase sweet corn with GE traits because possible consumer aversion to GE crops could negatively affect purchases of other products under their brand names (Bradford and Alston, 2004). Commercialized Bt sweet corn also has been engineered to resist glufosinate; however, glufosinate is not registered for use with Bt sweet corn because of concerns about consumer acceptance (Fennimore and Doohan, 2008).

Commercialized Genetically Engineered Crops Not Presently Available

Tomato.[a] The Flavr Savr tomato, developed by the company Calgene, was commercialized in 1994. The genetics of the tomato were engineered to slow the softening of the vegetable during ripening. The trait was developed in a tomato variety usually used for processing. However, a public opposition campaign against GE tomatoes caused some large processors to refuse to purchase the Flavr Savr variety for their products. In response, Calgene tried to sell the variety as a fresh-market tomato, but the vegetable bruised easily. That characteristic caused problems in production, transportation, and distribution. Furthermore, the Flavr Savr did not taste better than its cheaper competitors. Production of the variety was discontinued. Another GE tomato, developed for processing by the company Zeneca, was grown in California in the mid-1990s. Those tomatoes had a similar GE trait for delayed ripening and were processed into tomato paste for sale in the United Kingdom. However, consumer opposition to GE products caused Zeneca to discontinue the sale of the tomato paste in 1999.

Potato. A Bt potato resistant to the Colorado potato beetle was commercialized in 1995. Three years later, the technology developer, Monsanto, introduced a stacked variety that combined the Bt trait with virus resistance. Researchers found the Bt trait protected the potato from insect damage at all stages of the beetle's life (Perlak et al., 1993), and Monsanto scientists noted a large potential for reduction in the use of pesticides to treat insect and virus problems (Kaniewski and Thomas, 2004). However, Monsanto discontinued the sale of GE potatoes in 2001. The cultivars failed to capture more than 2–3 percent of the market for two reasons. First, a new insecticide that controlled the Colorado potato beetle and other pests came on the market at around the same time as GE potatoes; most farmers chose the insecticide over the GE trait (Nesbitt, 2005). Second, potato processors experienced a public-pressure campaign against the use of GE potatoes (Kilman, 2000; Kaniewski and Thomas, 2004). As food companies pledged to use non-GE potatoes in their products, farmers responded to processors' contracts for conventional varieties. Thus, although GE potatoes were technologically successful, they did not survive in the marketplace.

Alfalfa. Alfalfa is an important crop in the United States and is widely cultivated over a broad geographic range (USDA-NASS, 2009b). GE glyphosate-resistant alfalfa was commercialized in 2005, and about 198,000 acres was planted in 2006 (Weise, 2007). However, legal action over concerns about the risk of introgression of the transgene into nontransgenic alfalfa and the inability to mitigate this risk resulted in the termination of further seed sales and planting of glyphosate-resistant alfalfa (Charles, 2007) until USDA completed an environmental impact statement. That statement was released for public comment in December 2009.

---

[a]Adapted from Vogt and Parish (2001).

TABLE 1-1 Percentage of Soybean Acres in Genetically Engineered Soybean Varieties, by State and United States, 2000–2009

| | Herbicide-Resistant Soybean | | | | | | | | | |
| | 2000 | 2001 | 2002 | 2003 | 2004 | 2005 | 2006 | 2007 | 2008 | 2009 |
State	Percent of all soybean planted									
Arkansas	43	60	68	84	92	92	92	92	94	94
Illinois	44	64	71	77	81	81	87	88	87	90
Indiana	63	78	83	88	87	89	92	94	96	94
Iowa	59	73	75	84	89	91	91	94	95	94
Kansas	66	80	83	87	87	90	85	92	95	94
Michigan	50	59	72	73	75	76	81	87	84	83
Minnesota	46	63	71	79	82	83	88	92	91	92
Mississippi	48	63	80	89	93	96	96	96	97	94
Missouri	62	69	72	83	87	89	93	91	92	89
Nebraska	72	76	85	86	92	91	90	96	97	96
North Dakota	22	49	61	74	82	89	90	92	94	94
Ohio	48	64	73	74	76	77	82	87	89	83
South Dakota	68	80	89	91	95	95	93	97	97	98
Wisconsin	51	63	78	84	82	84	85	88	90	85
Other states[a]	54	64	70	76	82	84	86	86	87	87
United States	54	68	75	81	85	87	89	91	92	91

[a]Includes all other states in soybean estimating program.
SOURCE: USDA-NASS, 2001, 2003, 2005, 2007, 2009b.

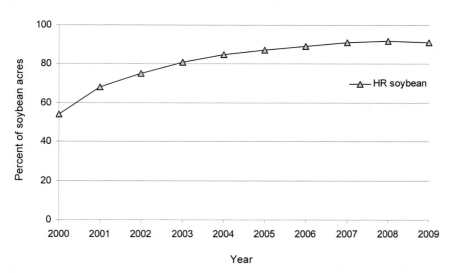

FIGURE 1-4 Herbicide-resistant soybean acreage trends nationwide.
SOURCE: USDA-NASS, 2001, 2003, 2005, 2007, 2009b.

Corn

The first GE variety of corn, which was commercialized in 1996, expressed a Bt toxin that targeted European corn borer, southwestern corn borer, and several other pests (see Table 1-2). GE corn with resistance to glyphosate was released in 1997, followed by a variety with resistance to glufosinate in the next year (Dill, 2005). An IR variety with a different Bt toxin to combat corn rootworm (*Diabrotica* spp.) was introduced in 2003.

Adoption of HR corn proved slower than that of soybean: Only 8 percent of the acreage was planted to HR corn in 2001 (Table 1-3, Figure 1-5). The low adoption rate of HR corn in 2001 was consistent among all U.S. regions. The narrow window of time for glyphosate application to be effective against early-season weed pressure in corn may have deterred farmer adoption (Tharp and Kells, 1999; Johnson et al., 2000; Gower et al., 2003; Kneževič et al., 2003; Dalley et al., 2004; Cox et al., 2005). Growers probably relied on traditional strategies for preemergence herbicide weed control rather than risk missing the glyphosate application window and ending up with weedier fields and reduced corn yields. Furthermore, lack of market access for HR corn to the European Union provided an added deterrent against early adoption of HR corn in the late 1990s and early 2000s.

TABLE 1-2 Insect Pests of Corn Targeted by Bt Varieties

Common Name	Latin Binomial
Primary Pest	
European corn borer	*Ostrinia nubilalis*
Southwestern corn borer	*Diatraea grandiosella*
Western corn rootworm	*Diabrotica virgifera virgifera*
Northern corn rootworm	*Diabrotica barberi*
Corn earworm	*Helicoverpa zea*
Fall armyworm	*Spodoptera frugiperda*
Black cutworm	*Agrotis ipsilon*
Secondary Pest	
Mexican corn rootworm	*Diabrotica virgifera zeae*
Southern cornstalk borer	*Diatraea crambidoides*
Stalk borer	*Papaipema nebris*
Lesser cornstalk borer	*Elasmopalpus lignosellus*
Sugarcane borer	*Diatraea saccharalis*
Western bean cutworm	*Richia albicosta*

NOTE: This pest categorization does not describe specific pest pressures in different states or regions. For example, the sugarcane borer is a primary pest of corn in Louisiana.
SOURCE: US-EPA, 2009.

TABLE 1-3 Percentage of Corn Acres in Genetically Engineered Corn Varieties, by State and United States, 2000–2009

| | Insect-Resistant (Bt) Only | | | | | | | | | |
	2000	2001	2002	2003	2004	2005	2006	2007	2008	2009
State	Percent of all corn planted									
Illinois	13	12	18	23	26	25	24	19	13	10
Indiana	7	6	7	8	11	11	13	12	7	7
Iowa	23	25	31	33	36	35	32	22	16	14
Kansas	25	26	25	25	25	23	23	25	25	24
Michigan	8	8	12	18	15	15	16	19	15	13
Minnesota	28	25	29	31	35	33	28	26	19	23
Missouri	20	23	27	32	32	37	38	30	27	23
Nebraska	24	24	34	36	41	39	37	31	27	26
North Dakota[a]						21	29	29	24	22
Ohio	6	7	6	6	8	9	8	9	12	15
South Dakota	35	30	33	34	28	30	20	16	7	6
Texas[a]						21	27	22	20	21
Wisconsin	13	11	15	21	22	22	22	19	14	13
Other states[b]	10	11	14	17	19	19	20	20	20	20
United States	18	18	22	25	27	26	25	21	17	17

| | Stacked-Gene Varieties | | | | | | | | | |
	2000	2001	2002	2003	2004	2005	2006	2007	2008	2009
State	Percent of all corn planted									
Illinois	1	1	1	1	2	5	19	40	52	59
Indiana	*	*	*	1	2	4	12	30	55	55
Iowa	2	1	3	4	8	11	18	37	53	57
Kansas	1	1	2	5	5	10	12	21	35	38
Michigan	*	2	2	3	4	5	10	19	33	42
Minnesota	2	4	4	7	11	11	16	28	40	41
Missouri	2	1	2	1	4	6	7	13	22	37
Nebraska	2	2	4	5	6	12	15	25	35	42
North Dakota[a]						15	20	22	31	41
Ohio	*	*	*	*	1	2	5	20	37	35
South Dakota	2	3	10	17	21	22	34	43	58	65
Texas[a]						9	13	20	27	33
Wisconsin	1	1	2	2	2	6	10	22	35	37
Other states[b]	1	1	2	2	6	6	10	14	22	28
United States	1	1	2	4	6	9	15	28	40	46

*Less than 1%.
[a]Estimates published individually beginning in 2005.
[b]Includes all other states in corn estimating program.
SOURCE: USDA-NASS, 2001, 2003, 2005, 2007, 2009b.

Herbicide-Resistant Only

Percent of all corn planted

2000	2001	2002	2003	2004	2005	2006	2007	2008	2009
3	3	3	4	5	6	12	15	15	15
4	6	6	7	8	11	15	17	16	17
5	6	7	8	10	14	14	19	15	15
7	11	15	17	24	30	33	36	30	29
4	7	8	14	14	20	18	22	24	20
7	7	11	15	17	22	29	32	29	24
6	8	6	9	13	12	14	19	21	17
8	8	9	11	13	18	24	23	24	23
					39	34	37	34	30
3	4	3	3	4	7	13	12	17	17
11	14	23	24	30	31	32	34	30	25
					42	37	37	31	30
4	6	9	9	14	18	18	23	26	27
6	8	12	17	21	19	25	33	32	30
6	7	9	11	14	17	21	24	23	22

All GE Varieties

Percent of all corn planted

2000	2001	2002	2003	2004	2005	2006	2007	2008	2009
17	16	22	28	33	36	55	74	80	84
11	12	13	16	21	26	40	59	78	79
30	32	41	45	54	60	64	78	84	86
33	38	43	47	54	63	68	82	90	91
12	17	22	35	33	40	44	60	72	75
37	36	44	53	63	66	73	86	88	88
28	32	34	42	49	55	59	62	70	77
34	34	46	52	60	69	76	79	86	91
					75	83	88	89	93
9	11	9	9	13	18	26	41	66	67
48	47	66	75	79	83	86	93	95	96
					72	77	79	78	84
18	18	26	32	38	46	50	64	75	77
17	20	27	36	46	44	55	67	74	78
25	26	34	40	47	52	61	73	80	85

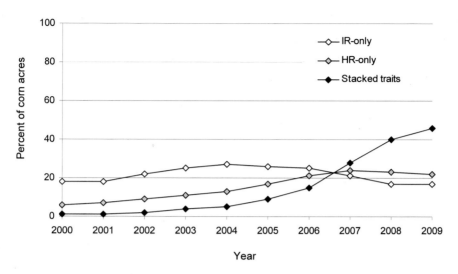

FIGURE 1-5 Genetically engineered corn acreage trends nationwide.
SOURCE: USDA-NASS, 2001, 2003, 2005, 2007, 2009b.

Variable insect pressure also delayed the adoption of IR corn, and this resulted in planting of only 19 percent of the acreage to IR corn in 2001 (Table 1-3, Figure 1-5). European corn borer is a key pest in the western Corn Belt region (Pilcher et al., 2002; Hyde et al., 2003; Mungai et al., 2005) but causes only a sporadic problem in the eastern Corn Belt region (Baute et al., 2002; Ma and Subedi, 2005; Cox et al., 2009). Consequently, IR corn acreage ranged from 23 to 30 percent in Iowa, Kansas, Minnesota, Missouri, Nebraska, and South Dakota but from 6 to 11 percent in Indiana, Michigan, Ohio, and Wisconsin in 2001 (Table 1-3). Farmers in regions without consistent corn borer infestations probably chose not to adopt IR corn.

In 2002, stacked hybrids were introduced, and this led to a further increase in acreage of GE corn. The increasing rate of adoption of stacked hybrids—2 percent in 2002 and 46 percent in 2009, with all major corn states above 30 percent (Table 1-3)—reflects the popularity of these traits and the lack of nonstacked GE traits in the seed marketplace. By 2009, 85 percent of U.S. corn acreage was planted with some type of GE seed; more than half these acres were in stacked varieties (Figure 1-5). In addition, by 2009, all major corn-growing states had GE acreage exceeding 70 percent except Ohio (67 percent); thus, adoption of IR corn is no longer region-specific (Table 1-3). Farmers' preference for multiple traits explains in part the lower rates of adoption of HR-only and IR-only varieties of corn compared with the rates of adoption of HR soybean (Figure 1-4).

Corn rootworm is a destructive and consistent pest in all regions of the United States that have continuous corn fields (and in some regions where corn is planted in fields after soybean). Bt corn for control of corn rootworm, especially western corn rootworm, has contributed to increased acreage of GE corn since its introduction in 2003 because growers preferred IR corn to the use of soil-applied insecticides or the use of insecticide and fungicide applied to seed at 1.25 mg of active ingredient[1] per seed. Bt corn hybrid seed for corn rootworm control is sold with only 0.25 mg of active ingredient per seed of insecticide and fungicide for control of secondary pests[2] and soil-borne pathogens. Growers can choose to add this feature for an additional cost to non-GE or HR corn hybrid seed. Thus, GE corn with the Bt trait for corn rootworm control and lower levels of seed-applied insecticide and fungicide substituted for the control tactics in continuous corn in the 1980s and 1990s of soil-applied insecticides for rootworm control and seed-applied products with higher toxicity[3] for secondary pest control, which growers had to manually apply to the seed. In-plant resistance for rootworm control with low levels of insecticide already applied to the seed by professional seed handlers for control of secondary corn pests is safer for the farmers who plant the crops and for the environment.

Cotton

Commercialized in 1996, IR cotton rapidly gained substantial market share because of its control of tobacco budworm, pink bollworm, and cotton bollworm (Table 1-4). GE glyphosate-resistant cotton, introduced in 1997, also proved popular with farmers because weed management has traditionally been more challenging in cotton than in many other field crops (Jost et al., 2008). The stacked Bt-glyphosate–resistant variety was introduced in 1997. By 2001, GE cotton had captured 69 percent of the acreage: 32 percent HR-only, 13 percent IR-only, and 24 percent stacked varieties (Table 1-5, Figure 1-6). Farmers in the southeast Cotton Belt adopted GE varieties more rapidly (78–91 percent in Arkansas, Georgia, Louisiana, Mississippi, and North Carolina) compared with those in Texas (49 percent) and California (40 percent), reflecting the lower insect pres-

---

[1]The *active ingredient* is the material in the pesticide that is biologically active. The active ingredient is typically mixed with other materials to improve the pesticide's handling, storage, and application properties.

[2]Examples include click beetles (*Alaus oculatus*), scarab beetles (*Scarabaeus sacer*), seed corn maggot (*Delia platura*), and wireworms (*Melanotus* spp).

[3]Examples include O,O-diethyl 0-2-isopropyl-6-methyl(pyrimidine-4-yl) phosphorothioate (commonly marketed as Diazinon); N-trichloromethylthio-4-cyclohexene-1,2-dicarboximide (Captan); and gamma-hexachlorocyclohexane (Lindane).

TABLE 1-4 Insect Pests of Cotton Targeted by Bt Varieties

Common Name	Latin Name
Primary Pest	
Cotton bollworm	*Helicoverpa zea*
Tobacco budworm	*Heliothis virescens*
Pink bollworm	*Pectinophora gossypiella*
Secondary Pest	
Salt marsh caterpillar	*Estigmene acrea*
Cotton leaf perforator	*Bucculatrix thurberiella*
Soybean looper	*Pseudoplusia includens*
Beet armyworm	*Spodoptera exigua*
Fall armyworm	*Spodoptera frugiperda*
Yellowstriped armyworm	*Spodoptera ornithogalli*
European corn borer	*Ostrinia nubilalis*

NOTE: This pest categorization does not describe specific pest pressures in different states or regions. For example, cotton bollworm and tobacco budworm are minor pests of cotton in Arizona.
SOURCE: US-EPA, 2009.

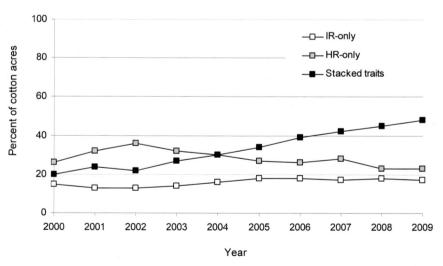

FIGURE 1-6 Genetically engineered cotton acreage trends nationwide.
SOURCE: USDA-NASS, 2001, 2003, 2005, 2007, 2009b.

sure in the latter two states, especially California (only 11 percent IR and 2 percent stacked varieties).

A new HR variety introduced in 2006 provided growers with a wider window for glyphosate application and the possibility of using higher glyphosate dosages (Mills et al., 2008). At around the same time, IR cotton with two Bt endotoxins was commercialized and offered improved control of cotton bollworm, increased protection against such secondary pests as beet armyworm and soybean looper, and advantages in resistance management (Mills et al., 2008; Siebert et al., 2008). The introduction of the improved traits alone or in stacked cultivars contributed to the increase in GE cotton to 88 percent in 2009: 23 percent HR-only, 17 percent IR-only, and 48 percent stacked (Table 1-5). As in 2001, farmers in the southeastern states had a higher adoption rate of GE cotton in 2009 (91 percent or greater) than Texas (81 percent) and California (73 percent). Pink bollworm and cotton bollworm are not major insect pests in California, so adoption of IR cotton (8 percent) and stacked varieties (11 percent) are particularly low; HR cotton (54 percent) makes up most of GE cotton in California.

An Early Portrait of Farmers Who Adopt Genetically Engineered Crops

A study of cotton farmers' planting decisions in four southeastern U.S. states in 1996 and 1997 provided early evidence on the various factors that influenced the choice to adopt transgenic cotton (Marra et al., 2001). The growers were asked about their human capital (stock of knowledge and ability), farm-specific characteristics, reasons for adopting or not adopting Bt cotton in 1996, and the pest-control regimens that they used on both their conventional and their Bt cotton acres (if applicable), including amounts and types of insecticides applied and their costs. Comparing the farmers' actions on fields planted to Bt and non-Bt cotton on the same farm controlled for variation in management, land quality, and machinery complement.

The study found that one measure of human capital that was associated with a higher likelihood of adopting Bt cotton was experience (number of years of growing cotton). The age of the farmer was not significant. The propensity to adopt because of higher profit potential of genetic-engineering technology—related to higher yields, decreased costs, or both—was also affirmed by the responding farmers. They reported higher yields on their Bt acres than on their non-Bt acres (6.58 lb/acre more on fields with Bt cotton than those without in the Upper South and 16.43 lb/acre more in the Lower South) and large reductions in pesticide costs in both regions (about $6.00/acre less for fields with Bt cotton in the

TABLE 1-5 Percentage of Cotton Acres in Genetically Engineered Upland Cotton Varieties, by State and United States, 2000–2009

| | Insect-Resistant (Bt) Only | | | | | | | | | |
	2000	2001	2002	2003	2004	2005	2006	2007	2008	2009
State	Percent of all upland cotton planted									
Alabama[a]						10	10	10	18	13
Arkansas	33	21	27	24	34	42	28	32	30	28
California	3	11	6	9	6	8	9	4	7	8
Georgia	18	13	8	14	13	29	19	17	19	20
Louisiana	37	30	27	30	26	21	13	17	19	20
Mississippi	29	10	19	15	16	14	7	16	19	14
Missouri[a]						20	32	13	12	18
North Carolina	11	9	14	16	18	17	19	13	19	15
Tennessee[a]						13	16	10	10	7
Texas	7	8	7	8	10	14	18	16	16	15
Other states[b]	17	18	19	18	22	18	21	27	22	24
United States	15	13	13	14	16	18	18	17	18	17

| | Stacked-Gene Varieties | | | | | | | | | |
	2000	2001	2002	2003	2004	2005	2006	2007	2008	2009
State	Percent of all upland cotton planted									
Alabama[a]						54	60	60	65	60
Arkansas	14	28	26	46	45	42	45	47	64	64
California	4	2	1	3	7	5	8	6	8	11
Georgia	32	29	30	47	58	55	64	68	73	70
Louisiana	30	47	49	46	60	64	68	68	73	63
Mississippi	36	61	47	61	58	59	69	62	66	63
Missouri[a]						16	25	23	19	51
North Carolina	36	38	45	48	46	54	60	64	62	68
Tennessee[a]						75	67	71	73	80
Texas	6	6	4	6	8	14	18	28	31	35
Other states[b]	36	33	32	38	45	46	45	42	48	49
United States	20	24	22	27	30	34	39	42	45	48

[a]Estimates published individually beginning in 2005.
[b]Includes all other states in upland cotton estimating program.
SOURCE: USDA-NASS, 2001, 2003, 2005, 2007, 2009b.

Upper South and about $10.00/acre less in the Lower South). Similarly, farmers who had previously experienced a high degree of pest infestation or pest resistance to currently used pesticides were more inclined to grow Bt cotton. Adopters reported higher past boll damage (7 percent higher on average compared with nonadopters) and higher incidence of past pest resistance to conventional insecticides (31 percent reported pest resistance

Herbicide-Resistant Only

2000	2001	2002	2003	2004	2005	2006	2007	2008	2009
Percent of all upland cotton planted									
					28	25	25	15	18
23	29	37	25	15	12	21	16	4	5
17	27	26	27	39	40	40	51	45	54
32	43	55	32	23	11	13	10	5	7
13	14	9	15	7	10	13	11	6	10
13	15	22	16	23	23	22	19	13	16
					59	40	63	68	29
29	37	27	29	27	24	19	16	14	13
					8	10	17	14	10
33	35	40	39	40	35	34	36	31	31
21	33	35	32	24	24	24	20	20	17
26	32	36	32	30	27	26	28	23	23

All GE Varieties

2000	2001	2002	2003	2004	2005	2006	2007	2008	2009
Percent of all upland cotton planted									
					92	95	95	98	91
70	78	90	95	94	96	94	95	98	97
24	40	33	39	52	53	57	61	60	73
82	85	93	93	94	95	96	95	97	97
80	91	85	91	93	95	94	96	98	93
78	86	88	92	97	96	98	97	98	93
					95	97	99	99	98
76	84	86	93	91	95	98	93	95	96
					96	93	98	97	97
46	49	51	53	58	63	70	80	78	81
74	84	86	88	91	88	90	89	90	90
61	69	71	73	76	79	83	87	86	88

compared with 18 percent of nonadopters in the combined sample). Those findings on human capital, yields, and the influence of pest problems are in accord with the explanations for adoption put forth by the diffusion and threshold theories.

Farm characteristics can also play a role in the decision to adopt a new technology. If the technology requires a high initial investment (such as for new machinery), farmers with more acres over which to spread the fixed costs might be more likely to adopt. Although the production technology itself is considered to be scale-neutral (i.e., the technology should not have differential impacts based on the size of the farm operation into which it is adopted), adopters in the study in both regions tended to have much larger farms and to farm more cotton acres than nonadopters; this supports the idea that the costs of learning may not be scale-neutral and thus that there is a possibility that differential farm-level social impacts have been associated with the adoption of GE crops (explained further in Chapter 4).

In 2001, farmers in Indiana, Illinois, Iowa, Minnesota, and Nebraska were surveyed to analyze the differences between adopters and nonadopters in farm and farmer characteristics (Wilson et al., 2005). The responses revealed that farmers growing corn on farms of less than 160 acres planted a greater percentage to GE corn for European corn borer control (54.5 percent) than farmers growing corn on farms of over 520 acres (39.2 percent). The same small–large differential held for aerial application of an insecticide (73.8 percent of farmers with less than 160 acres versus 57.3 percent with more than 520 acres); this suggests that smaller farmers place greater reliance on both chemical and GE controls of European corn borer than larger farmers. Just over one-fifth of the farmers (21.1 percent) reported a yield increase with the use of transgenic corn for European corn borer in all five states, from 11.2 percent in Indiana to 29.9 percent in Minnesota; 2.8 percent reported a yield decrease; and the rest reported no change in yield or that they did not know if there was a change or not. The surveyed farmers' greatest concerns were the ability to sell GE grain (59.3 percent), a market-access factor, and the additional technology fee (57.3 percent), a production-input factor that affected profits. Finally, the responding farmers indicated that a reduction in exposure to chemical insecticide (69.9 percent of the farmers), a personal health concern, and a reduction in insecticides in the environment (68.5 percent), a personal value, were the primary benefits of transgenic corn.

A more recent study of GE-crop adoption pertains to soybean (Marra et al., 2004). Table 1-6 presents the average total number of operated acres, the proportion of operated acres owned, age, education, and income (by category) for the different classes of adopters with the results of pairwise t-test results. The t-test results show that adopters (both partial and full) in this survey tended to be younger and operated more acres than nonadopters. Income, education, and percentage of operated acres the farmer owned do not show statistically significant differences among classes of adopters (Marra et al., 2004).

The importance of social networks in influencing patterns of adoption of GE crops has been highlighted in another recent study of the

TABLE 1-6 National Soybean Survey Descriptive Statistics by
Adoption Category

Farm Characteristics	Nonadopters	Partial Adopters	Full Adopters
Total Operated Acres	916.9^a	1237.5^b	1193.4^b
(N)	(44)	(66)	(136)
Proportion of Acres Owned	0.6^a	0.5^a	0.5^a
(N)	(59)	(78)	(167)
Year Born	1944.7^a	1947.8^b	1946.3^b
(N)	(54)	(72)	(150)
Years of Formal Education	13.2^a	13.7^a	13.3^a
(N)	(44)	(62)	(131)
Total Income (by category)	3.3^a	2.8^a	3.0^a
(N)	(34)	(49)	(103)

NOTE: If a superscript letter is different, the mean for this class of adopters is statistically
significantly different from the others in that category.
NOTE: Income categories ranged from 1= <\$50,000/year to 5 = >\$500,000/year.
SOURCE: Marra et al., 2004.

adoption of Bt corn in the Midwest. It described how farmers, whom
the author of the study termed *reflexive producers*, negotiate between the
advice and claims of experts, who do not farm, and local forms of knowledge that are conveyed by members of farmer networks. The study found
that farmers' determination of whether pest problems that require the
use of Bt corn exist depended more on local than on expert knowledge
(Kaup, 2008).

DETERRENTS TO GENETICALLY ENGINEERED
TRAIT DEVELOPMENT IN OTHER CROPS

Soybean, corn, and cotton represent a substantial number of acres
planted in the United States, but they do not reflect the diversity of
American agriculture. GE varieties have not been developed by private
firms for most U.S. crops, in part because the small markets for these
crops will not generate sufficient returns on the necessary investment in
research, technology commercialization, and marketing infrastructure.
Furthermore, concern about selling food with GE-derived ingredients
in some markets and the resistance of some grower organizations have
limited the commercial application of genetic-engineering technology to
just a few crops.

Market Conditions Influencing the Commercialization
of Genetically Engineered Varieties

Most research in and development of GE crops are conducted by private firms. Private companies must produce profits for their shareholders, so the marketability of a crop plays a determining role in decisions as to which GE crops are brought to commercialization. Market size, trait value, regulatory costs, environmental concerns, and technology access influence biotechnology firms' decisions to develop and sell GE seeds.

The market for seeds must be large enough to warrant the investment in commercialization. If markets are too small or are characterized by farmers with low ability to pay for the technology, the benefits to firms are too low to induce them to introduce GE varieties. That is one of the reasons that specialty crops have largely been overlooked in genetic engineering. The VR papaya, for example, was developed through public research. In addition, the number of researchers in these types of crops is considerably smaller and the marketing infrastructure less extensive than for soybean, corn, and cotton. That lack of resources, the diversity of species, the relatively short marketing season, and the small number of planted acres combine to deter private-sector investment in genetic-engineering technology for specialty crops (Bradford and Alston, 2004). To collect sufficient returns, firms instead invest in widely grown crops that have long storage life and that have year-round marketing potential. That generally means that farmers growing such crops have access to genetic-engineering technology, whereas the option is not available to farmers growing specialty crops or crops that are not widely grown in the United States.

The cost of regulatory compliance to ensure that GE crops do not pose unacceptable food safety and environmental risks has become an important component of the overall cost of new biotechnologies (Kalaitzandonakes et al., 2007). These costs may have contributed to limiting the development of GE minor crops, as was the case with pesticide development during the 1970–1990 period. As Ollinger and Fernandez-Cornejo (1995) found, "pesticide regulations have encouraged firms to focus their chemical pesticide research on pesticides for larger crop markets and abandon pesticide development for smaller crop markets." Obtaining regulatory clearance of GE crops in the United States is a long process, and the cost per crop can be very high. Furthermore, for crops with wild, weedy relatives (e.g., wheat), the potential for gene flow raises their environmental risk and expense (see "Gene Flow and Genetically Engineered Crops" in Chapter 2). Large private firms have concluded that investment in less widely grown crops does not generate adequate returns to justify the development and regulatory cost of bringing them to market.

Research and development in genetic-engineering technology have been stimulated by the development of patent protection for GE organisms. Changes in intellectual-property rights (IPR) law in the 1970s and 1980s are largely responsible for creating a profitable environment for biotechnology research. However, that protection may also create constraints on the development of GE varieties of more crops. Companies that control the patents may be unwilling to provide licenses or offer licenses at affordable prices to public-sector researchers or other companies that would like to develop seeds for smaller markets. A similar restriction may occur when university scientists patent genetic material that becomes essential for development of GE crops by other university scientists. Thus, the mechanism that generated the incentives to develop and commercialize genetic engineering may limit its applicability to most crops (Alston, 2004). The influence of IPR on the commercialization of genetically engineered crops will be discussed further in Chapter 4.

Marketing decisions are also influenced by perceived consumer acceptance of GE products. If technology providers have reason to believe that a GE crop will not be purchased by consumers, the technology will not be commercialized regardless of the potential benefits of the technology to producers. Indeed, a product may even be decommercialized if consumer avoidance, or the fear of it, is high enough. For example, consumer concerns and competing pest-control products caused the GE potato to be discontinued (see Box 1-2). The perceived potential loss of markets has also postponed the commercialization of GE wheat (this is covered further in Chapter 4). Consumers appear to be more accepting of products that are further removed from direct consumption, although additional research is needed in this regard (Tenbült et al., 2008). Thus, companies have been more willing to invest in corn and soybean, which are used primarily for animal feed and processed products, and cotton, a fiber crop. Even though wheat and rice are grains (like corn), are widely planted, and have a considerable storage life, their proximity to the consumer in the food supply chain has contributed to additional pressures on the private sector, which may explain firms' wariness to introduce genetic-engineering technology into them (Wisner, 2006).

Resistance to Genetic-Engineering Technology in Organic Agriculture

As outlined above, genetic-engineering technology is not available to farmers of most crops. However, some producers have chosen not to adopt the technology regardless of its accessibility. That attitude is typified by organic production in the United States.

As American agricultural practices incorporated greater use of synthetic chemicals in the 1950s and 1960s, organic production gained popularity as an alternative farming system. By the 1980s, the organic movement was large enough to justify the establishment of national certification standards. The proliferation of standards, inconsistency in labeling, difficulty in marketing, and inability to police violators of standards prompted organic groups to push for passage of the Organic Foods Production Act (OFPA) of 1990 (Rowson, 1998). The OFPA authorized a National Organic Program (NOP) in the U.S. Department of Agriculture (USDA) to define organic farming practices and acceptable inputs. The act established an advisory group, the National Organic Standards Board (NOSB), to provide recommendations to USDA on the structure and guidelines of the NOP. The NOSB viewed GE organisms as inconsistent with the principles of organic agriculture and recommended their exclusion (Vos, 2000). Opponents of genetic-engineering technology in organic production raised concerns about food safety and environmental effects. They also argued that organic agriculture is based on a set of values that places a high priority on "naturalness" (Verhoog et al., 2003), a criterion that in their view genetic engineering did not meet.

The proposed rule that was issued in 1997 deemed GE seeds permissible in organic agriculture; subsequently, USDA received a record number of public comments, almost entirely in objection to the proposal (Rowson, 1998). In response to the opposition, USDA rewrote the standards. When the NOP final rule went into effect in 2001, GE plants were not considered to be compliant with standards of organic agriculture (Johnson, 2008).

FROM ADOPTION TO IMPACT

The assessment framework described earlier in this chapter spans all the qualitative dimensions necessary to evaluate the potential sustainability of genetic-engineering technology. Therefore, this report's structure covers environmental, economic, and social changes, and the following chapters report progress and conclusions in these realms.

Environmental Effects

The landscape-level environmental effects of GE crops, both potential improvements and risks, did not receive extensive study when such crops were first planted widely (Wolfenbarger and Phifer, 2000; Ervin et al., 2000; Marvier, 2002). Since then, many studies on nontarget effects, including further studies requested by the U.S. Environmental Protection Agency, have accumulated. Other studies and analyses have related adoption of GE crops to changes in pesticide regimens and tillage practices. However, longitudinal data are still needed to better understand the effects of changes in farm management on environmental sustainability, such as on water quality or on resistance to glyphosate in weeds. Comprehensive evidence on other environmental dimensions—such as some aspects of soil quality, biodiversity, water quality and quantity, and air-quality effects—is also sparse. The environmental effects of farmers' adoption of genetic-engineering technology are discussed in Chapter 2.

Economic Effects

The economic effects of genetic-engineering technology in agriculture, which are addressed in Chapter 3, stem from effects on crop yields; the market returns received for the products; reductions or increases in production inputs and their prices, such as the costs of GE seeds and pesticides; and such other effects as labor savings that permit more off-farm work or that result in changes in yield risk. Those effects have received considerable study, particularly in the early stages of adoption of GE crops. However, recent information is sparse even though new GE varieties continue to be introduced. Less farm-level economic analysis has been conducted, perhaps because of the near dominance of the technologies in soybean, cotton, and corn production, because serious production or environmental problems have not surfaced, and because there is less interest for conducting additional research in a well-studied arena. More extensive studies of some economic effects, such as those on yield, have been conducted more recently in developing countries than in mature markets such as the United States.

Social Effects

The social effects of the adoption or nonadoption of genetic-engineering technology have not been studied as extensively as those attributed to previous waves of technological development in agriculture, even though earlier studies demonstrated that revolutionary agricultural technologies generally have substantial impacts at the farm or community level (Berardi, 1981; DuPuis and Geisler, 1988; Buttel et al., 1990) and that there was a high expectation that genetic-engineering technology would also have substantive and varied social impacts (Pimentel et al., 1989). It is thus surprising that there has been relatively little research on the ethical and socioeconomic effects of the adoption of agricultural biotechnology at the farm or community level (e.g., Buttel, 2005). A few studies have explored the economic effects of structural changes (integration and concentration) in the seed and agrichemical industries (Hayenga, 1998; Brennan et al., 1999; Fulton and Giannakas, 2001; Fernandez-Cornejo and Schimmelpfennig, 2004; Fernandez-Cornejo and Just, 2007). However, though the issue of how farmers might be socially impacted by the increasing integration of seed and chemical companies was first raised more than 20 years ago (Hansen et al., 1986), the organizations responsible for conducting or sponsoring research on the effects of genetic-engineering technology have generally fallen short of promoting the comprehensive and rigorous assessment of the possible social and ethical effects of GE-crop adoption. That responsibility rests not only with federal agencies (Kinchy et al., 2008) but with state governments, universities, nongovernment organizations, and the private for-profit sector. The absence of such research reduces our ability to document what the effects of the adoption of genetic-engineering technology have been on farm numbers and structure, community socioeconomic development, and the health and well-being of farm managers, family members, and hired farm laborers. A particularly significant question that has not been adequately assessed is whether the adoption of GE crops has exacerbated, alleviated, or had a neutral effect on the steady decline of farm numbers and the vitality of rural communities often associated with the industrialization of U.S. agricultural production. Because of the comparative dearth of empirical research findings on the social impacts of GE-crop adoption in the United States, we offer in Chapter 4 a discussion of the potential effects of the introduction of genetic-engineering technologies on farming-system dynamics in the form of testable hypotheses and piece together the ancillary literature on documented social effects, such as legal disputes.

CONCLUSION

Genetic-engineering technology has been built on centuries of plant-breeding experiments, research, and technology development. Commercialized applications have focused on pest management, primarily through resistance to the herbicide glyphosate and the incorporation of endotoxins that are lethal to some insect pests. Those traits have provided farmers of soybean, corn, and cotton with additional tools for combating pests. The popularity of GE crops is evidenced by their widespread adoption by farmers. In the following three chapters, we examine how their adoption has changed or reinforced farming practices and what implications the changes have for environmental, economic, and social sustainability at the farm level. At the close, we identify remaining challenges and opportunities for GE crops in the United States and draw conclusions and recommendations for increasing their contributions to farm sustainability.

REFERENCES

Alston, J.M. 2004. Horticultural biotechnology faces significant economic and market barriers. *California Agriculture* 58(2):80–88.

Alston, J.M., J.A. Hyde, M.C. Marra, and P.D. Mitchell. 2002. An ex ante analysis of the benefits from the adoption of corn rootworm resistant transgenic corn technology. *AgBioForum* 5(3):71–84.

Andersen, N.S., H.R. Siegismund, V. Meyer, and R.B. Jørgensen. 2005. Low level of gene flow from cultivated beets (*Beta vulgaris* L. ssp. *vulgaris*) into Danish populations of sea beet (*Beta vulgaris* L. ssp. *maritima* (L.) *Arcangeli*). *Molecular Ecology* 14(5):1391–1405.

Arce, A., and T.K. Marsden. 1993. The social construction of international food: A new research agenda. *Economic Geography* 69(3):293–311.

Baute, T.S., M.K. Sears, and A.W. Schaafsma. 2002. Use of transgenic *Bacillus thuringiensis* Berliner corn hybrids to determine the direct economic impact of the European corn borer (Lepidoptera: Crambidae) on field corn in eastern Canada. *Journal of Economic Entomology* 95(1):57–64.

Berardi, G.M. 1981. Socio–economic consequences of agricultural mechanization in the United States: Needed redirections for mechanization research. *Rural Sociology* 46(3):483–504.

Bonanno, A. 1991. The restructuring of the agricultural and food system: Social and economic equity in the reshaping of the agrarian question and the food question. *Agriculture and Human Values* 8(4):72–82.

Boyd, W. 2003. Wonderful potencies? Deep structure and the problem of monopoly in agricultural biotechnology. In *Engineering trouble: Biotechnology and its discontents*. eds. R. Schurman and D.D. Kelso, pp. 24–62. Berkeley: University of California Press.

Bradford, K.J., and J.M. Alston. 2004. Sidebar: Diversity of horticultural biotech crops contributes to market hurdles. *California Agriculture* 58(2):84–85.

Bradley, K.W., N.H. Monnig, T.R. Legleiter, and J.D. Wait. 2007. Influence of glyphosate tank-mix combinations and application timings on weed control and yield in glyphosate-resistant soybean. *Crop Management*. Available online at http://www.plantmanagementnetwork. org/pub/cm/research/2007/tank/. Accessed April 7, 2009.

Brennan, M.F., C.E. Pray, and A. Courtmanche. 1999. Impact of industry concentration on innovation in the U.S. plant biotech industry. Paper presented at the Transitions in agbiotech: Economics of strategy and policy NE-165 conference (Washington, DC, June 24–25, 1999).

Brown, D.C.W., and T.A. Thorpe. 1995. Crop improvement through tissue culture. *World Journal of Microbiology and Biotechnology* 11(4):409–415.

Buccola, S., D. Ervin, and H. Yang. 2009. Research choice and finance in university bioscience. *Southern Economic Journal* 75(4): 1238–1255.

Busch, L., and A. Juska. 1997. Beyond political economy: Actor networks and the globalization of agriculture. *Review of International Political Economy* 4(4):688–708.

Buttel, F.H. 2005. The environmental and post–environmental politics of genetically modified crops and foods. *Environmental Politics* 14(3):309–323.

Buttel, F.H., O.F. Larson, and G.W. Gillespie Jr. 1990. *The sociology of agriculture*. New York: Greenwood Press.

Calflora. 2009. Information on California plants for education, research and conservation. The Calflora Database. Available online at http://www.calflora.org/cgi-bin/specieslist. cgi?orderby=taxon&where-genus=Beta&ttime=1251561416&ttime=1251561417. Accessed August 29, 2009.

Cerdeira, A.L., and S.O. Duke. 2006. The current status and environmental impacts of glyphosate–resistant crops: A review. *Journal of Environmental Quality* 35(5):1633–1658.

Charles, D. 2007. Environmental regulation. U.S. courts say transgenic crops need tighter scrutiny. *Science* 315(5815):1069.

Conway, G. 1998. *The doubly green revolution: Food for all in the twenty–first century*. Ithaca, NY: Comstock Pub. Associates.

Corrigan, K.A., and R.G. Harvey. 2000. Glyphosate with and without residual herbicides in no–till glyphosate–resistant soybean (*Glycine max*). *Weed Technology* 14(3):569–577.

Cox, W.J., J. Hanchar, and E. Shields. 2009. Stacked corn hybrids show inconsistent yield and economic responses in New York. *Agronomy Journal* 101(6):1530–1537.

Cox, W.J., R.R. Hahn, P.J. Stachowski, and J.H. Cherney. 2005. Weed interference and glyphosate timing affect corn forage yield and quality. *Agronomy Journal* 97(3):847–853.

Crosby, A.W. 2003. *The Columbian exchange: Biological and cultural consequences of 1492.* 30th anniversary ed. Westport, CT: Praeger.

Dalley, C.D., J.J. Kells, and K.A. Renner. 2004. Effect of glyphosate application timing and row spacing on weed growth in corn (*Zea mays*) and soybean (*Glycine max*). *Weed Technology* 18(1):177–182.

Dill, G.M. 2005. Glyphosate-resistant crops: History, status and future. *Pest Management Science* 61(3):219–224.

Dill, G.M., C.A. CaJacob, and S.R. Padgette. 2008. Glyphosate–resistant crops: Adoption, use and future considerations. *Pest Management Science* 64(4):326–331.

Duke, S.O. 2005. Taking stock of herbicide-resistant crops ten years after introduction. *Pest Management Science* 61(3):211–218.

Duke, S.O., and S.B. Powles. 2008. Glyphosate–resistant weeds and crops. *Pest Management Science* 64(4):317–318.

DuPuis, E.M., and C. Geisler. 1988. Biotechnology and the small farm. *BioScience* 38(6):406–411.

Ervin, D.E., S.S. Batie, R. Welsh, C.L. Carpenter, J.I. Fern, N.J. Richman, and M.A. Schulz. 2000. Transgenic crops: An environmental assessment. Arlington, VA: Winrock International. Available online at http://www.winrock.org/wallace/wallacecenter/documents/transgenic.pdf. Accessed August 1, 2009.

Feder, G., R.E. Just, and D. Zilberman. 1985. Adoption of agricultural innovations in developing countries: A survey. *Economic Development and Cultural Change* 33(2):255–298.

Fennimore, S.A., and D.J. Doohan. 2008. The challenges of specialty crop weed control, future directions. *Weed Technology* 22(2):364–372.

Fernandez-Cornejo, J. 2004. The seed industry in U.S. agriculture: An exploration of data and information on crop seed markets, regulation, industry structure, and research and development. Agriculture Information Bulletin No. 786. U.S. Department of Agriculture–Economic Research Service. Washington, DC. Available online at http://www.ers.usda.gov/publications/aib786/aib786.pdf. Accessed May 26, 2009.

Fernandez-Cornejo, J., and D. Schimmelpfennig. 2004. Have seed industry changes affected research effort? *Amber Waves* 2(1):14–19.

Fernandez-Cornejo, J., and R.E. Just. 2007. Researchability of modern agricultural input markets and growing concentration. *American Journal of Agricultural Economics* 89(5):1269–1275.

Fernandez-Cornejo, J., R. Nehring, E.N. Sinha, A. Grube, and A. Vialou. 2009. Assessing recent trends in pesticide use in U.S. agriculture. Paper presented at the 2009 Annual Meeting of the Agricultural and Applied Economics Association (Milwaukee, WI, July 26–28, 2009). Available online at http://ageconresearch.umn.edu/handle/49271. Accessed June 16, 2009.

Fischer, A.J., A.J. Arnold, and M. Gibbs. 1996. Information and speed of innovation adoption. *American Journal of Agricultural Economics* 78(4):1073–1081.

Food, Agriculture, Conservation, and Trade Act of 1990, 101 P.L. 624; 104 Stat. 3359; 16 U.S.C., Subtitle A, section 1603.

Foster, A.D., and M.R. Rosenzweig. 1995. Learning by doing and learning from others: Human capital and technical change in agriculture. *Journal of Political Economy* 103(6):1176–1209.

Friedland, W.H. 2002. Agriculture and rurality: Beginning the "final separation"? *Rural Sociology* 67(3):350–371.

Fuchs, M., and D. Gonsalves. 2008. Safety of virus–resistant transgenic plants two decades after their introduction: Lessons from realistic field risk assessment studies. *Annual Review of Phytopathology* 45:173–202.

Fulton, M., and K. Giannakas. 2001. Agricultural biotechnology and industry structure. *AgBioForum* 4(2):137–151.

Gianessi, L.P. 2005. Economic and herbicide use impacts of glyphosate-resistant crops. *Pest Management Science* 61(3):241–245.

Gower, S.A., M.M. Loux, J. Cardina, S.K. Harrison, P.L. Sprankle, N.J. Probst, T.T. Bauman, W. Bugg, W.S. Curran, R.S. Currie, R.G. Harvey, W.G. Johnson, J.J. Kells, M.D.K. Owen, D.L. Regehr, C.H. Slack, M. Spaur, C.L. Sprague, M. VanGessel, and B.G. Young. 2003. Effect of postemergence glyphosate application timing on weed control and grain yield in glyphosate-resistant corn: Results of a 2-yr multistate study. *Weed Technology* 17(4):821–828.

Hansen, M., L. Busch, J. Burkhardt, W.B. Lacy, and L.R. Lacy. 1986. Plant breeding and biotechnology: New technologies raise important social questions. *BioScience* 36(1):29–39.

Hayenga, M.L. 1998. Structural change in the biotech seed and chemical industrial complex. *AgBioForum* 1(2):43–55.

Hill, T. 2009. Personal communication to the Committee on the Impact of Biotechnology on Farm-Level Economics and Sustainability. February 26. Washington, DC.

Howatt, K.A. 2008. Personal communication to M.D.K. Owen.

Hubbell, B.J., M.C. Marra, and G.A. Carlson. 2000. Estimating the demand for a new technology: Bt cotton and insecticide policies. *American Journal of Agricultural Economics* 82(1):118–132.

Hyde, J.A., M.A. Martin, P.V. Preckel, L.L. Buschman, C.R. Edwards, P.E. Sloderbeck, and R.A. Higgins. 2003. The value of Bt corn in southwest Kansas: A Monte Carlo simulation approach. *Journal of Agricultural and Resource Economics* 28(1):15–33.

James, C. 2009. *Global status of commercialized biotech/GM crops: 2009*. ISAAA Brief No. 41 ed. The International Service for the Acquisition of Agri-biotech Applications. Ithaca, NY.

Johnson, R. 2008. Organic agriculture in the United States: Program and policy issues. RL31595. Congressional Research Service–Resources Science and Industry Division. Washington, DC. Available online at http://www.ncseonline.org/NLE/CRSreports/06Oct/RL31595.pdf. Accessed February 3, 2009.

Johnson, W.G., P.R. Bradley, S.E. Hart, M.L. Buesinger, and R.E. Massey. 2000. Efficacy and economics of weed management in glyphosate-resistant corn (*Zea mays*). *Weed Technology* 14(1):57–65.

Jost, P., D. Shurley, S. Culpepper, P. Roberts, R. Nichols, J. Reeves, and S. Anthony. 2008. Economic comparison of transgenic and nontransgenic cotton production systems in Georgia. *Agronomy Journal* 100(1):42–51.

Just, D.R., S.A. Wolf, S. Wu, and D. Zilberman. 2002. Consumption of economic information in agriculture. *American Journal of Agricultural Economics* 84(1):39–52.

Kalaitzandonakes, N., J.M. Alston, and K.J. Bradford. 2007. Compliance costs for regulatory approval of new biotech crops. *Nature Biotechnology* 25(5):509–511.

Kaniewski, W.K., and P.E. Thomas. 2004. The potato story. *AgBioForum* 7(1&2):41–46. Available online at http://www.agbioforum.missouri.edu/v7n12/v7n12a08-kaniewski.htm. Accessed May 21, 2009.

Kaup, B.Z. 2008. The reflexive producer: The influence of farmer knowledge upon the use of Bt corn. *Rural Sociology* 73(1):62–81.

Kilman, S. 2000. Companies feel pressure from worries over bio-crops; Monsanto's potato losing appeal with fast-food firms. *The San Diego Union-Tribune*, April 29. p. C-1, Business section.

Kinchy, A.J., D.L. Kleinman, and R. Autry. 2008. Against free markets, against science? Regulating the socio-economic effects of biotechnology. *Rural Sociology* 73:147–179.

Kloppenburg, J.R. 2004. *First the seed: The political economy of plant biotechnology, 1492–2000*. 2nd ed. Madison, WI: University of Wisconsin Press.

Kneževič, S.Z., S.P. Evans, and M. Mainz. 2003. Yield penalty due to delayed weed control in corn and soybean. *Crop Management*. Available online at http://www.plantmanagementnetwork.org/pub/cm/research/2003/delay/. Accessed April 8, 2009.

Kniss, A.R., R.G. Wilson, A.R. Martin, P.A. Burgener, and D.M. Feuz. 2004. Economic evaluation of glyphosate-resistant and conventional sugar beet. *Weed Technology* 18:388–396.

Lewontin, R. 1990. The political economy of agricultural research: The case of hybrid corn. In *Agroecology*. eds. C.R. Carroll, J.H. Vandermeer, and P. Rosset, pp. 613–626. New York: McGraw-Hill.

Lichtenberg, E., and D. Zilberman. 1986a. The econometrics of damage control—why specification matters. *American Journal of Agricultural Economics* 68(2):261–273.

———. 1986b. The welfare economics of price supports in U.S. agriculture. *American Economic Review* 76(5):1135–1141.

Liu, W., and M. Tollenaar. 2009. Physiological mechanisms underlying heterosis for shade tolerance in maize. *Crop Science* 49(5):1817–1826.

Lynch, R.E., B.R. Wiseman, D. Plaisted, and D. Warnick. 1999. Evaluation of transgenic sweet corn hybrids expressing Cry1A(b) toxin for resistance to corn earworm and fall armyworm (Lepidoptera: Noctuidae). *Journal of Economic Entomology* 92:246–252.

Ma, B.L., and K.D. Subedi. 2005. Development, yield, grain moisture and nitrogen uptake of Bt corn hybrids and their conventional near-isolines. *Field Crops Research* 93(2-3):199–211.

Marra, M.C., B.J. Hubbell, and G.A. Carlson. 2001. Information quality, technology depreciation, and Bt cotton adoption in the Southeast. *Journal of Agricultural and Resource Economics* 26(1):158–175.

Marra, M.C., D.J. Pannell, and A. Abadi Ghadim. 2003. The economics of risk, uncertainty and learning in the adoption of new agricultural technologies: Where are we on the learning curve? *Agricultural Systems* 75(2-3):215–234.

Marra, M.C., N.E. Piggott, and G.A. Carlson. 2004. The net benefits, including convenience, of Roundup Ready® soybeans: Results from a national survey. Technical Bulletin No. 2004–3. NSF Center for Integrated Pest Management. Raleigh, NC. Available online at http://cipm.ncsu.edu/cipmpubs/marra_soybeans.pdf. Accessed April 8, 2009.

Marvier, M. 2002. Improving risk assessment for nontarget safety of transgenic crops. *Ecological Applications* 12(4):1119–1124.

Mason, M. 2009. Personal communication via email to K. Laney and the Committee on the Impact of Biotechnology on Farm-Level Economics and Sustainability. December 1. Washington, DC.

Mazoyer, M., and L. Roudart. 2006. *A history of world agriculture: From the neolithic age to the current crisis*. Translated by J.H. Membrez. London: Earthscan.

Mills, C.I., C.W. Bednarz, G.L. Ritchie, and J.R. Whitaker. 2008. Yield, quality, and fruit distribution in Bollgard/Roundup Ready and Bollgard II/Roundup Ready flex cottons. *Agronomy Journal* 100(1):35–41.

Morgan, K., T. Marsden, and J. Murdoch. 2006. *Worlds of food: Place, power, and provenance in the food chain*. New York: Oxford University Press.

Mouzelis, P.N. 1976. Capitalism and the development of agriculture. *Journal of Peasant Studies* 3:483–492.

Mulugeta, D., and C.M. Boerboom. 2000. Critical time of weed removal in glyphosate–resistant *Glycine max*. *Weed Science* 48(1):35–42.

Mungai, N.W., P.P. Motavalli, K.A. Nelson, and R.J. Kremer. 2005. Differences in yields, residue composition and N mineralization dynamics of Bt and non-Bt maize. *Nutrient Cycling in Agroecosystems* 73(1):101–109.

Nesbitt, T.C. 2005. GE foods in the market. Ithaca, NY: Cornell University Cooperative Extension.

Ollinger, M., and J. Fernandez-Cornejo. 1995. Regulation, innovation, and market structure in the U.S. pesticide industry. Agricultural Economic Report No. 719. U.S. Department of Agriculture–Economic Research Service. Washington, DC.

Perlak, F.J., T.B. Stone, Y.M. Muskopf, L.J. Petersen, G.B. Parker, S.A. McPherson, J. Wyman, S. Love, G. Reed, D. Biever, and D.A. Fischhoff. 1993. Genetically improved potatoes: Protection from damage by Colorado potato beetles. *Plant Molecular Biology* 22(2):313–321.

Piggott, N.E., and M.C. Marra. 2007. The net gain to cotton farmers of a natural refuge plan for Bollgard II® cotton. *AgBioForum* 10(1):1–10.

———. 2008. Biotechnology adoption over time in the presence of non-pecuniary characteristics that directly affect utility: A derived demand approach. *AgBioForum* 11(1):58–70.

Pilcher, C.D., M.E. Rice, R.A. Higgins, K.L. Steffey, R.L. Hellmich, J. Witkowski, D. Calvin, K.R. Ostlie, and M. Gray. 2002. Biotechnology and the European corn borer: Measuring historical farmer perceptions and adoption of transgenic Bt corn as a pest management strategy. *Journal of Economic Entomology* 95(5):878–892.

Pimentel, D., M.S. Hunter, J.A. Lagro, R.A. Efroymson, J.C. Landers, F.T. Mervis, C.A. McCarthy, and A.E. Boyd. 1989. Benefits and risks of genetic engineering in agriculture. *BioScience* 39(9):606–614.

Rogers, E.M. 2003. *Diffusion of innovations*. New York: Free Press.

Rowson, G. 1998. Organic foods and the proposed federal certification and labeling program. 98-264 ENR. Congressional Research Service–Environment and Natural Resources Division. Washington, DC.

Roy, B.A. 2004. Rounding up the costs and benefits of herbicide use. *Proceedings of the National Academy of Sciences of the United States of America* 101(39):13974–13975.

Scursoni, J., F. Forcella, J. Gunsolus, M. Owen, R. Oliver, R. Smeda, and R. Vidrine. 2006. Weed diversity and soybean yield with glyphosate management along a north-south transect in the United States. *Weed Science* 54(4):713–719.

Siebert, M.W., S. Nolting, B.R. Leonard, L.B. Braxton, J.N. All, J.W. Van Duyn, J.R. Bradley, J. Bacheler, and R.M. Huckaba. 2008. Efficacy of transgenic cotton expressing CrylAc and CrylF insecticidal protein against heliothines (Lepidoptera: Noctuidae). *Journal of Economic Entomology* 101(6):1950–1959.

Stachler, J.M. 2009. Personal communication to M.D.K. Owen.

Stewart, C.N., Jr., M.D. Halfhill, and S.I. Warwick. 2003. Transgene introgression from genetically modified crops to their wild relatives. *Nature Reviews Genetics* 4(10):806–817.

Stokstad, E. 2008. GM papaya takes on ringspot virus and wins. *Science* 320(5875):472.

Storstad, O., and H. Bjørkhaug. 2003. Foundations of production and consumption of organic food in Norway: Common attitudes among farmers and consumers? *Agriculture and Human Values* 20(2):151–163.

Sunding, D., and D. Zilberman. 2001. The agricultural innovation process: Research and technology adoption in a changing agricultural sector. In *Handbook of agricultural economics*, Vol. 1, Part 1. eds. B.L. Gardner and G.C. Raussers, pp. 207–261. Amsterdam: Elsevier.

Tenbült, P., N.K. De Vries, G. van Breukelen, E. Dreezens, and C. Martijn. 2008. Acceptance of genetically modified foods: The relation between technology and evaluation. *Appetite* 51(1):129–136.

Tharp, B.E., and J.J. Kells. 1999. Influence of herbicide application rate, timing, and interrow cultivation on weed control and corn (*Zea mays*) yield in glufosinate-resistant and glyphosate-resistant corn. *Weed Technology* 13(4):807–813.

US-EPA (U.S. Environmental Protection Agency). 2009. Pesticides. Washington, DC. Available online at http://www.epa.gov/pesticides/. Accessed January 8, 2010.

USDA-NASS (U.S. Department of Agriculture–National Agricultural Statistics Service). 2001. Acreage. June 29. Cr Pr 2-5 (6-01). Washington, DC. Available online at http://usda.mannlib.cornell.edu/usda/nass/Acre//2000s/2001/Acre-06-29-2001.pdf. Accessed April 14, 2009.

———. 2003. Acreage. June 30. Cr Pr 2-5 (6-03). Washington, DC. Available online at http://usda.mannlib.cornell.edu/usda/nass/Acre//2000s/2003/Acre-06-30-2003.pdf. Accessed April 14, 2009.

———. 2005. Acreage. June 30. Cr Pr 2-5 (6-05). Washington, DC. Available online at http://usda.mannlib.cornell.edu/usda/nass/Acre//2000s/2005/Acre-06-30-2005.pdf. Accessed April 14, 2009.

———. 2007. Acreage. June 29. Cr Pr 2-5 (6-07). Washington, DC. Available online at http://usda.mannlib.cornell.edu/usda/nass/Acre//2000s/2007/Acre-06-29-2007.pdf. Accessed April 14, 2009.

———. 2008. Acreage. June 30. Cr Pr 2-5 (6-08). Washington, DC. Available online at http://usda.mannlib.cornell.edu/usda/current/Acre/Acre-06-30-2009.pdf. Accessed April 14, 2009.

———. 2009a. Quick stats (agricultural statistics database). Washington, DC. Available online at www.nass.usda.gov/Data_and_Statistics/index.asp. Accessed September 14, 2009.

———. 2009b. Acreage. June 30. Cr Pr 2-5 (6-09). Washington, DC. Available online at http://usda.mannlib.cornell.edu/usda/current/Acre/Acre-06-30-2009.pdf. Accessed November 24, 2009.

Vasil, I.K. 2008. A short history of plant biotechnology. *Phytochemistry Reviews* 7(3):387–394.

Vavilov, N.I. 1951. The origins, varieties, immunity and breeding of cultivated plants. *Chronica Botanica* 13:1–366.

Verhoog, H., M. Matze, E.L. van Bueren, and T. Baars. 2003. The role of the concept of the natural (naturalness) in organic farming. *Journal of Agricultural and Environmental Ethics* 16(1):29–49.

Vogt, D.U., and M. Parish. 2001. Food biotechnology in the United States: Science, regulation, and issues. RL30198. Congressional Research Service. Washington, DC. Available online at http://ncseonline.org/NLE/CRSreports/science/st-41.pdf. Accessed October 21, 2009.

Vos, T. 2000. Visions of the middle landscape: Organic farming and the politics of nature. *Agriculture and Human Values* 17(3):245–256.

Weise, E. 2007. Effects of genetically engineered alfalfa cultivate a debate. *USA Today*, February 15. p. 10D, Life section. Available online at http://www.usatoday.com/news/health/2007-02-14-alfalfa_x.htm. Accessed June 4, 2009.

Wiesbrook, M.L., W.G. Johnson, S.E. Hart, P.R. Bradley, and L.M. Wax. 2001. Comparison of weed management systems in narrow-row, glyphosate- and glufosinate-resistant soybean (*Glycine max*). *Weed Technology* 15(1):122–128.

Wilson, T.A., M.E. Rice, J.J. Tollefson, and C.D. Pilcher. 2005. Transgenic corn for control of the European corn borer and corn rootworms: A survey of midwestern farmers' practices and perceptions. *Journal of Economic Entomology* 98(2):237–247.

Wisner, R. 2006. Potential market impacts from commercializing Round-Up Ready® wheat. September. Western Organization of Resource Councils. Billings, MT. Available online at http://www.worc.org/userfiles/file/Wisner-Market%20Risks-Update-2006.pdf. Accessed July 5, 2009.

Wolf, S.A., D.R. Just, and D. Zilberman. 2001. Between data and decisions: The organization of agricultural economic information systems. *Research Policy* 30(1):121–141.

Wolfenbarger, L.L., and P.R. Phifer. 2000. The ecological risks and benefits of genetically engineered plants. *Science* 290(5499):2088–2093.

2

Environmental Impacts of Genetically Engineered Crops at the Farm Level

The environmental impacts of planting genetically engineered (GE) crops occur within the context of agriculture's general contribution to environmental change. Agriculture has historically converted biologically diverse natural grasslands, wetlands, and native forests into less diverse agroecosystems to produce food, feed, and fiber. Effects on the environment depend on the intensity of cultivation over time and space; the inputs applied, including water, fertilizer, and pesticides; and the management of inputs, crop residue, and tillage. With 18 percent of the land area in the United States planted to crops and another 26 percent devoted to pastures (FAO, 2008), the huge scale of these impacts becomes obvious. In general, tillage, crop monoculture, fertilizers, and pesticide use often have adverse effects on soil, water, and biodiversity. Agriculture is the leading cause of water-quality impairment in the United States (USDA-ERS, 2006). No-tillage systems, crop rotations, integrated pest management, and other environmentally friendly management practices may ameliorate some of the adverse impacts, but the tradeoff between agricultural production and the environment remains. With agricultural lands approaching 50 percent of U.S. land, developing more ecologically and environmentally sound agricultural management practices for crops, soil, and water is a central challenge for the future (Hanson et al., 2008). Against that backdrop, we evaluate the impact of GE crops on the environmental sustainability of U.S. farms.

This chapter examines the changes in farm practices that have accompanied the adoption of GE crops and the evidence on how such adoption affects the environment. It addresses impacts at the individual farm level and also at the landscape level, given that impacts from individual farms accumulate and affect other farms and their access to communal natural resources in the region. The use of GE crops has altered farmers' agronomic practices, such as tillage, herbicides, and insecticides; these alterations have implications for environmental sustainability both on and off the farm, which are evaluated to the extent possible at this point in time (Box 2-1). In particular, we examine the effects of the adoption of GE crops on soil quality, biodiversity, and water quality.

ENVIRONMENTAL IMPACTS OF HERBICIDE-RESISTANT CROPS

The adoption of herbicide-resistant (HR) crops has affected the types and number of herbicides and the amount of active ingredient applied to soybean, corn, and cotton. This section first examines the substitution of glyphosate for other herbicides that has taken place and how the use of HR crops has interacted with tillage practices. It then assesses ecological effects of those changes on soil quality, water quality, arthropod biodiversity, and weed communities. Lastly, the implications for weed management in cropping systems with HR crops are considered, especially for systems in which glyphosate-resistant weeds evolve.

Herbicide Substitution

A higher proportion of herbicide-resistant GE soybean has been planted than of any other GE crop in the United States. Adoption has exceeded 90 percent of the acres planted to soybean by U.S. farmers (Figure 2-1). HR cotton acreage reached 71 percent in 2009 (Figure 2-2), while planted HR corn acres were 68 percent that year (Figure 2-3). The HR crops planted thus far have altered the mix of herbicides used in cropping systems and allowed the substitution of glyphosate for other herbicides.[1] Figures 2-1 through 2-3 summarize the trends in the use of glyphosate and other herbicides on soybean, cotton, and corn (expressed in pounds of active ingredient per planted acre of these crops) and the adoption of HR soybean, cotton, and corn (Fernandez-Cornejo et al., 2009). It is important to recognize that, depending on the metrics used, the substitution of glyphosate for other herbicides has resulted in the use of fewer alternative herbicides by growers of HR crops. However, glyphosate is often applied in higher doses and with greater frequency than the herbicides it replaced. Thus, the actual amount of active ingredi-

---

[1]A list of herbicides for which glyphosate is a common substitute can be found in Appendix A.

BOX 2-1
Limitations to Evaluating the Magnitude
of Environmental Effects

Although environmental risk assessment is conducted for all GE varieties before regulatory approval, in some cases the absence of environmental monitoring at the landscape level prevents calculating the magnitude of effects (e.g., water quality) following commercialization. Where monitoring data on agricultural practices are available (e.g., tillage practices, pesticide use), simple correlations of the adoption rates with trends in agricultural practices do not capture the complexity required to quantify the magnitude of any environmental effect. The lack of spatially explicit data linking the use of GE crops with data-monitoring agricultural practices stymies any accurate calculation of the magnitude of environmental effects at national or even regional levels (NRC, 2002). Environmental consequences of agricultural practices can vary greatly at a subregional scale. For example, the adoption of a herbicide-resistant crop may facilitate use of no-till practices, but the environmental effects of no-till practices depend on existing soil texture, structure, and erosion potential for each individual farm. Though models may exist to quantify soil retention given erosion potential, what amount of retention can be attributed to HR crops requires two additional calculations:

1. Quantifying to what extent HR crops caused the adoption of conservation tillage practices, given that there is a two-way relationship, and

2. Spatially linking the adoption of HR crops with data on the occurrence of Highly Erodible Land, something not feasible without spatially-explicit data.

Similarly, weed and pest control measures fluctuate from year to year and crop to crop, as have the choices of active ingredients. Determining the extent to which adoption of GE crops replaced specific pesticides over time requires incorporating a suite of factors, such as changes in pest pressure or pest-management strategies (e.g., see footnote on boll weevil eradication program), tillage practices, technology, and public policy (e.g., pesticide regulation, government programs) (Fernandez-Cornejo et al., 2009). Spatial data on the evolution of weed resistance are also lacking, thus preventing any calculation of environmental consequences of the declining effectiveness of glyphosate with glyphosate-resistant crops.

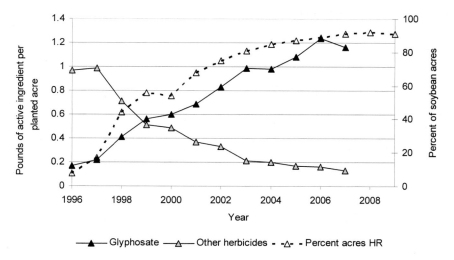

FIGURE 2-1 Application of herbicide to soybean and percentage of acres of herbicide-resistant soybean.
NOTE: The strong correlation between the rising percentage of HR soybean acres planted over time, the increased applications of glyphosate, and the decreased use of other herbicides suggests but does not confirm causation between these variables.
SOURCE: USDA-NASS, 2001, 2003, 2005, 2007, 2009a, 2009b; Fernandez-Cornejo et al., 2009.

ents (glyphosate and other herbicides) applied per acre actually increased from 1996 to 2007 in soybean (Figure 2-1) and cotton (Figure 2-2) but decreased over the same period in corn (Figure 2-3).

Glyphosate is reported to be more environmentally benign than the herbicides that it has replaced (Fernandez-Cornejo and McBride, 2002; Cerdeira and Duke, 2006). It binds to soil rapidly (preventing leaching), it is biodegraded by soil bacteria, and it has a very low toxicity to mammals, birds, and fish (Malik et al., 1989). Glyphosate can be detected in the soil for a relatively short period of time compared to many other herbicides, but is essentially biologically unavailable (Wauchope et al., 1992). Formulations that contain the surfactant polyoxyethylene amine can be toxic to some amphibians at environmentally expected concentrations and may affect aquatic organisms under some environmental conditions (Folmar et al., 1979; Tsui and Chu, 2003; Relyea and Jones, 2009); however, these formulations are labeled for terrestrial uses only with restrictions with respect to waterways. The greater use of postemergence glyphosate applications has been accompanied by modifications of agronomic practices, particularly

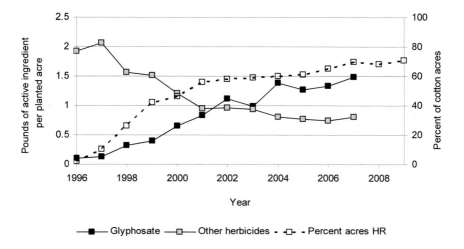

FIGURE 2-2 Application of herbicide to cotton and percentage of acres of herbicide-resistant cotton.
NOTE: The strong correlation between the rising percentage of HR cotton acres planted over time, the increased applications of glyphosate, and the decreased use of other herbicides suggests but does not confirm causation between these variables.
SOURCE: USDA-NASS, 2001, 2003, 2005, 2007, 2009a, 2009b; Fernandez-Cornejo et al., 2009.

with regards to weed management and tillage. The interactions of those practices have implications for environmental sustainability.

Tillage Practices

Tillage is one process used by farmers to prepare the soil before planting. In conventional tillage, all postharvest residue is plowed into the soil to prepare a clean seedbed for planting and to reduce the growth of weeds; in conservation tillage, at least 30 percent of the soil surface is left covered with crop residue after planting. In the 1970s and 1980s, innovations in cultivators and seeders enabled farmers to plant seeds at a reasonable cost with residue remaining on the field. Those developments encouraged the adoption of one form of conservation tillage called no-till, in which the soil and surface residue from the previously harvested crop are left undisturbed as the next crop is seeded directly into the soil without tillage. After soil-conservation policy was incorporated into the Food Security Act of 1985, conservation tillage accelerated in the 1990s. The

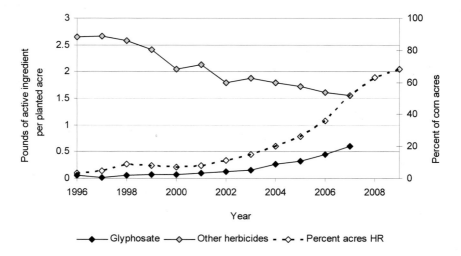

FIGURE 2-3 Application of herbicide to corn and percentage of herbicide-resistant corn.

NOTE: The strong correlation between the rising percentage of HR corn acres planted over time, the increased applications of glyphosate, and the decreased use of other herbicides suggests but does not confirm causation between these variables.

SOURCE: USDA-NASS, 2001, 2003, 2005, 2007, 2009a, 2009b; Fernandez-Cornejo et al., 2009.

introduction of HR soybean and cotton has supported the trend because the use of glyphosate allowed weeds to be controlled after crop emergence without the need for tillage to disrupt weed development before or after planting. Indeed, in the last 10 years, the use of conservation tillage has continued to increase, with the exception that it has remained constant in the case of corn (Figure 2-4).[2]

The adoption of conservation tillage practices by U.S. soybean growers increased from 51 percent of planted acres in 1996 to 63 percent in 2008, or an addition of 12 million acres. The adoption of no-till practices accounted for most of the increase and was used on 85 percent of these additional 12 million acres. Over the same time period, the acreage planted to soybean increased at most nine million acres. In cotton there was a doubling of the percentage of acres managed using conservation tillage from 1996 to 2008, and no-till is the predominant conservation tillage practice. Cotton acreage declined over the same time period. For

---

[2]More information on different types of tillage systems can be found in Appendix B.

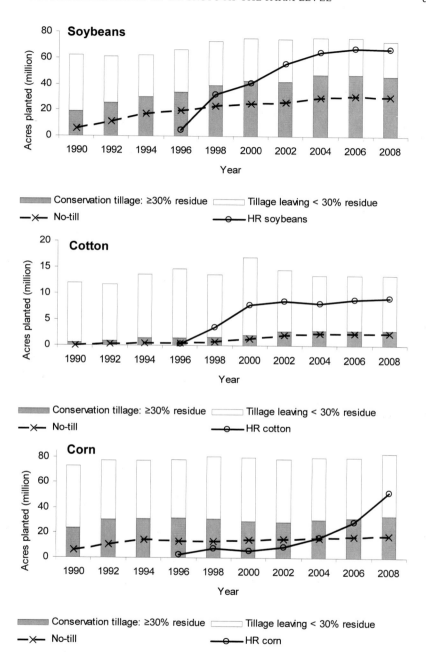

FIGURE 2-4 Trends in conservation tillage practices and no-till for soybean, cotton, and corn, and adoption of herbicide-resistant crops since their introduction in 1996.
SOURCE: CTIC, 2009; USDA-ERS, 2009.

corn between 1996 and 2008, an additional 4.8 million acres of corn were planted. At the same time, the use of conservation tillage practices remained at a fairly constant 40 percent of planted acreage. No-till practices increased by 4 percent over the same time period (4.3 million acres), but this was disproportionate relative to overall increases in conservation tillage practices (1.9 million acres), indicating that farmers converted from other conservation tillage practices to no-till.

According to U.S. Department of Agriculture (USDA) survey data for 1997, a larger share of acreages planted to HR soybean was managed with conservation tillage than was planted to conventional soybean (Fernandez-Cornejo and McBride, 2002)—about 60 percent versus about 40 percent (Figure 2-5). The difference in the use of no-till between adopters and nonadopters of HR soybean was even more pronounced: 40 percent of acres planted with HR soybean were under no-till, double the corresponding share of acres of non-GE soybean under no-till management practices (Fernandez-Cornejo and McBride, 2002).

From the perspective of farmer decision making, the availability of herbicide-resistance technology may affect the adoption of conservation tillage, and the use of conservation tillage may affect the decision to adopt HR crops. Several economists have tried to understand how closely the two decisions are linked. An econometric model developed to address the simultaneous nature of the decisions was used to determine the nature of the relationship between the adoption of GE crops with HR traits and no-till practices on the basis of 1997 national survey data on soybean farmers (Fernandez-Cornejo et al., 2003). Farmers using no-till were found to have a higher probability of adopting HR cultivars than farmers using conventional tillage, but using HR cultivars did not significantly affect no-till adoption. That result suggested that farmers already using no-till incorporated HR cultivars seamlessly into their weed-management program; but the commercialization of HR soybean did not seem to encourage the adoption of no-till, at least at the time of the survey.

More recently, however, Mensah (2007) found a two-way causal relationship on the basis of more recent data. Using a simultaneous-adoption model and 2002 survey data on soybean farmers, Mensah found that farmers who adopted no-till were more likely to adopt HR soybeans *and* that farmers who adopted herbicide-resistance technology were more likely to adopt no-till practices.

In the case of cotton, the evidence also points toward a two-way causal relationship. Roberts et al. (2006) evaluated the relationship between adoption of HR cotton and conservation tillage practices in Tennessee from 1992 to 2004. Using two methods,[3] they found that the adoption of

---

[3] An application of Bayes's theorem and a two-equation logit model.

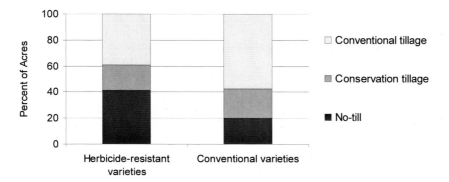

FIGURE 2-5 Soybean acreage under conventional tillage, conservation tillage, and no-till, 1997.
SOURCE: Adapted from Fernandez-Cornejo and McBride, 2002.

HR cotton increased the probability that farmers would adopt conservation tillage and conversely that farmers that had previously adopted conservation tillage practices were more likely to adopt HR cotton. Thus, the adoption of no-till and the adoption of HR cotton are complementary practices.

Kalaitzandonakes and Suntornpithug (2003) also studied the simultaneous adoption of HR and stacked cotton varieties and conservation tillage practices on the basis of farm-level data. They concluded that conservation tillage practices both encouraged the adoption of HR and stacked cotton varieties and were encouraged by their adoption. Using state-level data for 1997–2002 and using a simultaneous-equation econometric model, Frisvold et al. (2007) studied the diffusion of HR cotton and conservation tillage. They found strong complementarity between the two practices and rejected the null hypothesis that the diffusion of one is independent of the diffusion of the other. They also observed that an increase in the probability of adoption of HR cotton increased the probability of adoption of conservation tillage and vice versa.

Thus, most empirical evidence points to a two-way causal relationship between the adoption of HR crops and conservation tillage.[4] Farmers using conservation tillage practices are more likely to adopt HR crop varieties than those using conventional tillage, and those adopting HR crop varieties are more likely to change to conservation tillage practices than those who use non-HR cultivars. The analytical techniques used do

---

[4]Most published evidence is for the cases of soybean and cotton given that extensive adoption of HR corn is relatively more recent (HR corn adoption only exceeded 20 percent of corn acreage in 2005).

not reveal the relative strength of each causal linkage, so it is not clear which factor (adoption of HR varieties or use of conservation tillage) has a greater influence on the other.

Soil Quality

The relationship between the adoption of conservation tillage practices and the adoption of HR crops is relevant to farm sustainability because conservation tillage has fewer adverse environmental impacts than conventional tillage (reviewed by Uri et al., 1999). On the farm, conservation tillage reduces soil loss from erosion, increases water infiltration, and can improve soil quality and moisture retention (reviewed by Uri et al., 1999; Holland, 2004). Corn and soybean are grown in regions where highly erodible land is common, and conversion to conservation tillage for these crops results in substantial reduction in soil loss and wind erosion even on non-highly erodible land (Uri et al., 1999). Leaving more crop residue on fields strengthens nutrient cycling and increases soil organic matter, a key component of soil quality (reviewed by Blanco-Canqui and Lal, 2009). Soil organisms decompose plant residue, and this, in turn, cycles nutrients and improves soil structure. In general, soil organisms have greater abundance or biomass in no-till systems than in conventional tillage systems because soil is disturbed less (reviewed by Wardle, 1995; Kladivko, 2001; Liebig et al., 2004).

In addition to tillage, the use of herbicides can affect soil quality through their impact on soil organisms, so interpreting the effects of HR crops on soil quality requires an understanding of how tillage practices interact with herbicide use to influence the soil microorganism community. In laboratory studies, glyphosate can inhibit or stimulate microbial activity, depending on soil type and glyphosate formulation (Carlisle and Trevors, 1986, and references therein). Some microorganisms can use glyphosate as a substrate for metabolism (increased activity), whereas others are susceptible to the herbicide because they have an enzyme 5-enolpyruvyl-shikimate-3-phosphate synthase pathway that glyphosate inhibits. When species-level responses were measured, roots of glyphosate-resistant soybean and corn treated with glyphosate had significantly more colonies of the fungus *Fusarium* than did non-HR cultivars or HR cultivars not treated with glyphosate (Kremer and Means, 2009). In contrast, fluorescent *Pseudomonas* populations, an antagonist of fungal pathogens like *Fusarium*, were significantly lower in soybean that were both glyphosate resistant and treated with glyphosate compared to untreated HR cultivars or a non-HR cultivar treated with other herbicides (Kremer and Means, 2009). Those results indicate a change in the antagonistic relationship between *Fusarium* and *Pseudomonas* attributable to the formulation of glyphosate used. Whether magnitude of change in this antagonistic relationship would have consequences on soil quality of disease control was not a part of the study.

With respect to general microbial activity, three studies in the United States have detected no uniform changes in soil organism profiles in association with tillage or with the use of glyphosate on glyphosate-resistant cropping systems (Liphadzi et al., 2005; Weaver et al., 2007; Locke et al., 2008). Soil microorganisms in fields planted with glyphosate-resistant corn and soybean varieties were similar with and without tillage (Liphadzi et al., 2005). HR fields treated with glyphosate and non-GE fields treated with other herbicides were also similar in soil microbe activity (Liphadzi et al., 2005). On tilled, experimental plots of glyphosate-resistant soybean, transient changes in the soil microbial community were detected in the first few days after application of glyphosate compared to no application (Weaver et al., 2007), but the differences disappeared after 7 days. When there was continuous cotton cropping, soil quality did not differ between HR and non-HR systems. In contrast, soil under continuous HR-corn cropping contained more carbon and nitrogen than soil with non-HR corn (Locke et al., 2008), which would be considered a benign change. Differences in carbon and nitrogen contents could have been due to glyphosate use, but they were also probably influenced by changes in the detrital food web associated with the higher biomass of winter weeds in the HR-corn cropping system (Locke et al., 2008). Subtle differences in the structure of the soil microbial community were also detectable in those same experiments; the significance of the differences for soil quality were not discussed. Thus, species-level studies suggest that glyphosate can alter the microbial composition in the rhizosphere. General studies of the interaction of tillage and glyphosate use in HR crops have indicated transient benign effects of glyphosate and neutral, or in one case favorable, effects of conservation tillage on the soil communities in HR crops.

Water Quality

Conservation tillage practices can have off-farm benefits for water quality that are potentially more important than onsite productivity effects (Foster and Dabney, 1995). Because conservation tillage practices improve soil-water infiltration, the volume of runoff is less than when conventional tillage is used. Reduced tillage and no-till practices can improve water quality by reducing the amounts of sediments and sediment-associated chemicals in runoff from farm fields into surface water. Similarly, lower volumes of runoff can decrease the transport of soil nutrients and agricultural inputs, such as fertilizers and pesticides, although the decrease will vary with soil type, tillage practice, and nutrient or pesticide input.

For example, although the concentration of herbicide in runoff from no-till fields can be higher than when other conservation tillage practices are used, the total amount of herbicide in runoff may be similar because runoff volume is reduced (Fawcett et al., 1994; Locke and Bryson, 1997; Mickelson et al., 2001; Shipitalo and Owens, 2006; Zeimen et al., 2006). That phenomenon has been observed with the use of glyphosate in no-till fields (Shipitalo et al., 2008).

Studies have suggested that the use of glyphosate poses less risk to water quality than the use of other herbicides; this is attributable in part to the production systems typically used in GE crops and to the physical chemistry and relatively low toxicity of glyphosate (Estes et al., 2001; Wauchope et al., 2002; Peterson and Hulting, 2004). However, there are no regional-scale analyses of the effects of HR-crop adoption on water quality. One study conducted in a small Ohio watershed that compared herbicide runoff in HR and non-HR soybean fields found that the amount of glyphosate in the runoff was nearly one-seventh that of the herbicide metribuzin and about half that of alachlor, even though glyphosate was applied to soybean twice and alachlor and metribuzin once to soybean (Shipitalo et al., 2008). Those results are consistent with known characteristics of glyphosate, which strongly absorbs to soil and has a half-life in soil of 6–60 days, depending on soil characteristics. Microbial processes degrade glyphosate into two metabolites: sarcosine and aminomethylphosphonic acid (AMPA). Sarcosine degrades quickly to carbon dioxide and ammonia. AMPA is more persistent than glyphosate in the soil environment but is considered equally or less toxic (reviewed by Giesy et al., 2000). Numerous studies have documented the occurrence of glyphosate and AMPA in surface waters (Kolpin et al., 2006), but they have rarely been found in groundwater (Borggaard and Gimsing, 2008). Concentrations of glyphosate reported in surface water have not exceeded the maximum contaminant level (MCL) for drinking water set by the U.S. Environmental Protection Agency (EPA); in accordance with World Health Organization recommendations, MCLs have not been set for AMPA (WHO, 2005).

Shifts to conservation tillage attributable to the availability of HR crops have contributed to reductions in soil loss and probably in herbicide runoff. The magnitude and spatial distribution of the benefits are not precisely known, but the implications are that those are important environmental benefits of these cropping systems. However, as discussed later in this chapter (see "Other Shifts in Weed Communities"), some of the environmental benefits may be threatened in the future.

Arthropod Biodiversity

Changes in herbicide use and tillage practices like those accompanying the adoption of HR crops can affect such organisms as natural enemies of pests or pollinators, which provide ecological services to agriculture. Weeds provide an ecological disservice to farms by competing with crop plants for nutrients and light even at low population densities, but weeds can also support a broad array of nonpest species. Pollinators feed on nectar or use some weeds as hosts for their larval stage; weed species can be food for herbivores that in turn are preyed on by predators that also control pests of crops. In particular, more effective weed management could decrease the abundance of beneficial organisms, depending on the mobility of a species and how closely its resource base is associated with weed abundance. In contrast, the increase in no-till practices that leave more plant material undisturbed in fields may increase the resource base for beneficial insects.

Evidence indicates that the planting of HR cultivars does not consistently affect the weed diversity and abundance that support beneficial species. Whether a farmer used a GE crop or a conventional crop, better weed control has generally reduced the numbers of arthropods and other organisms in corn, sugar beet, and rapeseed fields (Hawes et al., 2003) and decreased the abundance of the predatory big-eyed bug (*Geocoris punctipes*) in soybean fields (Jackson et al., 2003; Jackson and Pitre, 2004). When HR crops improved weed management (decreased weeds), populations of natural enemies and pollinators decreased (Hawes et al., 2003). When conventional weed-management tactics (such as the use of the herbicide atrazine) were more effective at weed control on non-HR corn relative to HR corn, beneficial insect abundance was greater within the HR side of the field where more weeds occurred (Hawes et al., 2003). Subsequent analyses of these same data in more depth have revealed detailed associations between properties of the weed community and the accompanying arthropod food web (Hawes et al., 2009) and strengthened the conclusion that weed management accounts for the relationships observed. However, weed management was not the largest influence on the abundance of beneficial organisms. Rather, there were differences of a factor of 3–10 in abundance among different crops and between early and late in the growing season, compared with differences of a factor of 2 associated with weed management (Hawes et al., 2003).

Weed Biodiversity and Weed Shifts

Crop-production practices inevitably influence the composition of the weed community. Typically, only a few weed species are economically important in a particular crop-production system (Owen, 2001; Tuesca et al., 2001). When a production practice changes, for example, a change in herbicide, it may ultimately select for weed biotypes that are resistant to that herbicide (Baker, 1991). Other elements of production practices that have selective effects on the weed community include harvesting techniques, irrigation, fertilization, planting dates, soil amendments, and tillage (Hilgenfeld et al., 2004; Murphy and Lemerle, 2006; Owen, 2008).

The stronger the selective force of those practices (e.g., the level of disturbance caused by tillage), the more consistent the selective force (e.g., continuous planting of the same crop as opposed to annual crop rotations), and the simpler the selective force (e.g., the recurrent use of one herbicide), the greater the effect on the composition of the weed community (Owen, 2001). Changes in the kinds of weeds that are important locally are termed *weed shifts* (which implies changes in weed species composition) (Givnish, 2001); in the following discussion, weed shifts are the ecological process by which an initial weed community is replaced by a new community, including better-adapted species, in response to changes in agricultural practices. Weed shifts are generally followed by a period of stability, given the longevity of weed seeds in the soil, as long as the agricultural systems that resulted in the shift remain constant (Buhler, 1992; Buhler et al., 1997). They are a common and inevitable result of agriculture and are not unique to the adoption of HR crops, but it is essential to understand and manage them well if agriculture is to be productive and sustainable. Such shifts are particularly relevant for managing weeds in HR-crop systems, in which tillage practices and herbicide use both play major roles in shaping the weed community.

Herbicide Resistance in Weeds

The International Survey of Herbicide Resistant Weeds (ISHRW) provides a historical account and extensive list of weeds that have evolved resistance to herbicides (Heap, 2010). Although the ISHRW reflects the efforts of many weed scientists in reporting weed populations that have herbicide resistance, the voluntary basis of the contributions likely results in underestimation of the extent of resistance to herbicides, including glyphosate. The evolution of herbicide-resistant weeds is not unique to the herbicides for which HR traits exist. Currently, 195 species (115 dicots and 80 monocots) have evolved resistance to at least one of 19 herbicide mechanisms of action in at least 347 herbicide-resistant weed biotypes distributed over 340,000 fields (Heap, 2010).

Glyphosate, first commercialized in 1974, has been extensively used for weed control in perennial crops (fruits, trees, nuts, and vines), along roadsides and irrigation canal banks, and in urban areas and national parks (Powles, 2008). The first case of evolved resistance to glyphosate was reported in 1996 in rigid ryegrass (*Lolium rigidum*) (Powles et al.,

1998). The glyphosate-resistant population originated in an orchard in a large winter cropping region of southern Australia, where glyphosate had been used intensively for the control of rigid ryegrass for more than 15 years. Since the initial report, at least six other weed species have been reported as resistant to glyphosate in environments where glyphosate-resistant crops were not planted (Powles, 2008; Heap, 2010).

Emergence of Glyphosate-Resistant Weeds in Herbicide-Resistant Crop Fields

Ten species have evolved resistance to glyphosate independently in glyphosate-resistant crops over 14 years in the United States (from 1996 to 2010) (Heap, 2010). Gene flow between HR crops and closely related weed species does not explain the evolution of glyphosate resistance in U.S. fields because sexually compatible weeds are absent where corn, cotton, and soybean are grown in the United States. However, the nearly exclusive reliance on glyphosate for weed control, a practice accelerated by the widespread introduction of glyphosate-resistant crop varieties, has caused substantial changes in weed communities. The first report of glyphosate resistance associated with a GE glyphosate-resistant crop involved horseweed (*Conyza canadensis*) in Delaware (VanGessel, 2001); once resistance evolved, growers found it difficult to control this weed in no-till glyphosate-resistant soybean (VanGessel, 2001). Since the initial report in 2000, glyphosate-resistant populations of horseweed have been documented throughout the Mid-Atlantic, Mid-South, Mississippi Delta, and Midwest states (Heap, 2010). The weed grows particularly well in no-till production systems, producing a large number of wind-carried seeds that are dispersed over long distances (Buhler and Owen, 1997; Ozinga et al., 2004).

Subsequent to that discovery in 2000, other weed species have evolved resistance to glyphosate in glyphosate-resistant crops in the United States (Table 2-1). They include two species of pigweed, Palmer amaranth (*Amaranthus palmeri*) and waterhemp (*Amaranthus tuberculatus*),[5] which have become economically important in glyphosate-resistant cotton and soybean production (Zelaya and Owen, 2000, 2002; Culpepper, 2006; Culpepper and York, 2007; Legleiter and Bradley, 2008). Infested areas are increasing rapidly in the Southeast, the Mississippi Delta (Palmer amaranth as well as Johnsongrass, *Sorghum halepense*), and the Midwest (waterhemp) (Culpepper and York, 2007; Legleiter and Bradley, 2008). Glyphosate-resistant populations of giant ragweed (*Ambrosia trifida*) have been reported in several states (Leer, 2006), primarily in or adjacent to glyphosate-resistant soybean. Kochia (*Kochia scoparia*) with evolved resistance to glyphosate has recently been identified in Kansas (Heap, 2010).

---

[5]In some literature, *Amaranthus tuberculatus* is referred to as *Amaranthus rudis*.

TABLE 2-1 Weeds That Evolved Resistance to Glyphosate in Glyphosate-Resistant Crops in the United States

Species	Crop	Location	Acreage[a]
Amaranthus palmeri (Palmer amaranth)	Corn, cotton, soybean	Georgia, North Carolina, Arkansas, Tennessee, Mississippi	200,000–2,000,000
Amaranthus tuberculatus (waterhemp)	Corn, soybean	Missouri, Illinois, Kansas, Minnesota	1,200–11,000
Ambrosia artemisiifolia (common ragweed)	Soybean	Arkansas, Missouri, Kansas	<150
Ambrosia trifida (giant ragweed)	Cotton, soybean	Ohio, Arkansas, Indiana, Kansas, Minnesota, Tennessee	2,000–12,000
Conyza canadensis (horseweed)	Corn, cotton, soybean	14 states	> 2,000,000
Kochia scoparia (kochia)	Corn, soybean	Kansas	51–100
Lolium multiflorum (Italian ryegrass)	Cotton, soybean	Mississippi	1000–10,000
Sorghum halepense (Johnsongrass)	Soybean	Arkansas	Unknown

[a]Minimum and maximum acreages are based on expert judgments provided for each state. The estimates were summed and rounded to provide an assessment of the minimum and maximum acreages in the United States. These values indicate orders of magnitudes but do not provide precise information on abundance of resistant weeds.
SOURCE: Data from Heap, 2010.

Another weed, common lambsquarters (*Chenopodium album*) (Kniss et al., 2004, 2005; Schuster et al., 2007; Scursoni et al., 2007) may have also evolved glyphosate-resistant biotypes (Boerboom, 2005), but it has not yet appeared on the ISHRW list.

Other Shifts in Weed Communities

Factors other than the evolution of glyphosate resistance affect the composition of weed species in the field. Changes in the tillage system used in growing HR crops are probably the most important factor in promoting weed shifts because disturbance is a primary selective force (Buhler, 1992). In addition, weeds that escape glyphosate applications by germinating after the last application can have an advantage in glyphosate-resistant crops (Hilgenfeld et al., 2004; Owen and Zelaya, 2005; Puricelli and Tuesca, 2005; Scursoni et al., 2007; Wilson et al., 2007; Owen, 2008). Table 2-2 lists weed species that have been found to be naturally tolerant to the conditions prevalent in the fields where glyphosate-resistant crops are grown and have become more abundant after the widespread adoption of these crops. Shifts in local weed communities have been observed more frequently in glyphosate-resistant cotton and soybean than in glyphosate-resistant corn, probably because glyphosate-resistant cotton and soybean are more widely cultivated than glyphosate-resistant corn (Culpepper, 2006). However, where glyphosate-resistant corn and glyphosate-resistant soybean are commonly rotated (e.g., in the Midwest), strong selection pressure exists for the evolution of glyphosate-resistant weeds because the management tactics vary so little between the two crops.

Farmers' Response to Glyphosate Resistance in Weeds

The evolution of glyphosate resistance in some kinds of weeds and other weed shifts can diminish the technical and economic efficiency of weed control. However, because glyphosate allows producers to control a wide array of weeds conveniently and economically, they have been reluctant to stop using glyphosate-resistant crops and glyphosate when facing control problems arising from a few glyphosate-resistant or naturally glyphosate-tolerant weed species. For controlling problematic weeds, they prefer increasing the magnitude and frequency of glyphosate applications, using other herbicides in addition to glyphosate, or increasing their use of tillage.

For example, soybean growers in Delaware continued planting glyphosate-resistant soybean even in the presence of widespread glyphosate resistance in horseweed (Scott and VanGessel, 2007). Most producers addressed the problem by applying an herbicide with a different mode of

TABLE 2-2 Weeds Reported to Have Increased in Abundance in Glyphosate-Resistant Crops

Species	Crop	Location	Reference
Acalypha spp. (copperleaf)	Soybean	—	Owen and Zelaya, 2005; Culpepper, 2006
Amaranthus tuberculatus (waterhemp)	Soybean	—	Owen and Zelaya, 2005
Amaranthus palmeri (Palmer amaranth)	Cotton	—	Culpepper, 2006
Annual grasses	Cotton	—	Culpepper, 2006
Chenopodium album (common lambsquarters)	Soybean	Iowa, Minnesota	Owen, 2008
Commelina communis (Asiatic dayflower)	Cotton, soybean	Midwest, Midsouth, Southeast	Owen and Zelaya, 2005; Culpepper, 2006; Owen, 2008
Commelina benghalensis (tropical spiderwort)	Cotton	Southeast, Georgia	Owen, 2008; Mueller et al., 2005
Cyperus spp. (nutsedge)	Cotton	—	Culpepper, 2006
Equisetum arvense (field horsetail)	Herbicide-resistant crops	—	Owen, 2008
Oenothera biennis (evening primrose)	Herbicide-resistant crops	Iowa	Owen, 2008
Oenothera laciniata (cutleaf evening primrose)	Soybean	—	Culpepper, 2006
Pastinaca sativa (wild parsnip)	Herbicide-resistant crops	Iowa	Owen, 2008
Phytolacca americana (pokeweed)	Herbicide-resistant crops	—	Owen, 2008
Ipomoea spp. (annual morning glory)	Cotton	—	Culpepper, 2006

action, increasing the frequency of glyphosate applications, or using tillage before planting. Some 76 percent of growers estimated that resistance in horseweed increased their management costs by more than $2.02/acre, and 28 percent reported cost increases of over $8.09/acre (Scott and VanGessel, 2007). Similarly, a survey of 400 corn, soybean, and cotton producers in 17 states found that most would not limit the use of glyphosate-resistant crops when facing problematic glyphosate-resistant weeds (Foresman and Glasgow, 2008). Instead, producers planned to increase the rotation of herbicides, the use of tank-mixes, or the amount of tillage. They expected that additional measures for the control of glyphosate-resistant weeds would cost $13.90–16.30/acre (Foresman and Glasgow, 2008).

In an economic analysis of weed-management costs with a hypothetical reduction of control with glyphosate in three regions of the United States, the projected cost of new resistance-management practices for horseweed was $12.33/acre in a cotton–soybean–corn rotation in western Tennessee (Mueller et al., 2005). Additional costs were due to a shift from no-till to conventional tillage for cotton and the need for new preplant herbicides for soybean. The projected cost of new herbicide resistance-management practices for waterhemp was $17.91/acre in a corn–soybean rotation in southern Illinois; this cost resulted from use of different preemergence and postemergence herbicides for soybean (Mueller et al., 2005). For cotton grown in Georgia, the extra cost of controlling shifts in tropical spiderwort (*Commelina benghalensis*), a weed that is naturally tolerant to glyphosate, was predicted to be $14.91/acre; an additional herbicide application after cotton emergence explained this cost (Mueller et al., 2005).

Those studies indicate that the evolution of glyphosate resistance and weed shifts could lead to two important changes in practices: increased use of herbicides generally and reductions in conservation tillage (Mueller et al., 2005). Such changes would also increase weed-management costs and reduce producers' profits, and the environmental consequences of those practices, if they were widely adopted by producers of HR crops, would negate the environmental benefits previously achieved.

In summary, most glyphosate-resistant weeds in HR crops are of economic importance in row crops grown in the Southeast and Midwest. The number of weed species evolving resistance to glyphosate is growing (Figure 2-6), and the number of locations with glyphosate-resistant weeds is increasing at a greater rate, as more and more acreage is sprayed with glyphosate. Though the number of weeds with resistance to glyphosate is still small compared to other common herbicides,[6] the shift toward

---

[6]For example, 38 weeds have developed resistance to some acetyl-CoA carboxylase (ACCase) inhibitors, and resistance to some acetolactate synthase (ALS) inhibitors has been documented in 107 weed species worldwide.

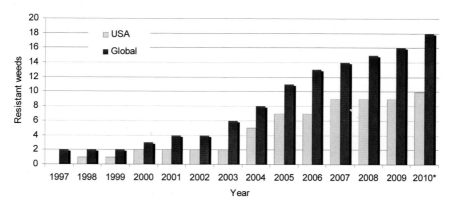

FIGURE 2-6 Number of weeds with evolved glyphosate resistance.
*Weed numbers are updated through March 2010.
SOURCE: Adapted from Heap, 2010.

glyphosate-resistant weed biotypes will probably become an even more important component of row-crop agriculture unless production practices (such as recurrent use of glyphosate) change dramatically (Gressel, 1996; Owen and Zelaya, 2005; Johnson et al., 2009).

Implications of Weed Shifts in Herbicide-Resistant Cropping Systems

As noted above, because the adoption of HR crops has facilitated an increase in conservation tillage and reduced the number of herbicides that growers use to control weeds, the selection pressures affecting weed communities has changed. Unsurprisingly, managing weeds through glyphosate applications to HR crops favors the evolution of glyphosate resistance in weeds occurring in these crop fields (Shaner, 2000; Mueller et al., 2005; Foresman and Glasgow, 2008; Powles, 2008). Addressing the problem of resistance—a problem not unique to HR crops—requires careful thought about management practices and other potential solutions based on a clear understanding of how genes that code for resistance are distributed throughout a population of a weed species.

Principles of Population Genetics Underlying Resistance in Weeds

Similar concepts have been used to understand the evolution of resistance to glyphosate in weeds and to the Bt toxin in insects. Population-genetic models and empirical data on factors that affect how resistance evolves have been applied to the management of both herbicide-resistant weeds and Bt-resistant insects (Jasieniuk et al., 1996; Werth et al., 2008). However, strategies for delaying the evolution and spread of resistance are not the same because there are important underlying differences in the population genetics of herbicide resistance and insect resistance.

Resistance to herbicides, and in particular to glyphosate, is often conferred by a single nuclear gene (Jasieniuk et al., 1996; Powles and Preston, 2006). Herbicide resistance in weeds is rarely recessive;[7] in all cases studied, resistance to glyphosate was additive to dominant, that is, individuals with a single resistance allele can survive applications of glyphosate (Jasieniuk et al., 1996; Zelaya et al., 2004; Powles and Preston, 2006; Zelaya et al., 2007; Neve, 2008). Furthermore, even if resistance is recessive in some weeds, many weeds are self-pollinating, so a recessive gene for resistance could become homozygous in only a few generations and thus confer resistance to all offspring (Gould, 1995; Jasieniuk et al., 1996). Some agronomically important weeds (such as pigweed) are dioecious (having separate male and female plants) and thus are cross-pollinated. They have demonstrated the ability to evolve resistance to glyphosate although the genetics of the process have not been described.

Finally, even though the seeds of some weed species can disperse over long distances (Shields et al., 2006), dispersal of viable pollen generally occurs over short distances (Jasieniuk et al., 1996; Roux et al., 2008). Therein lies an important difference between weeds and insects and hence the availability of strategies, such as refuges, to control weed resistance. For all the reasons described above, maintaining a refuge—an area where susceptible weeds are not exposed to glyphosate and would persist to interbreed with resistant biotypes—cannot be expected to lower the heritability of herbicide resistance in weeds as it lowers the heritability of Bt resistance in insects targeted by Bt crops (Jasieniuk et al., 1996). The refuge strategy for Bt-resistant insects is discussed later in this chapter (see "Evolution and Management of Insect Resistance").

Although the use of refuges cannot be expected to delay the evolution of glyphosate resistance in weeds, the spread of herbicide resistance can be delayed by reducing the selective differential (the difference in survival and other fitness traits) between individuals with and without resistance alleles (Gressel and Segel, 1990; Jasieniuk et al., 1996; Werth et al., 2008). That can be accomplished by using control practices that kill weeds that

---

[7]Plants carry two alleles (forms) of the same gene for glyphosate resistance. Each allele exerts influence on the nature of that trait; saying that resistance is recessive means that one allele is not sufficient to confer resistance. Offspring that inherit one allele with the resistant trait and one without will not be resistant, while offspring that inherit two of the same form of the allele that confers resistance will be resistant.

have the resistance alleles. For example, the use of tank-mixes that contain two or more herbicides with different modes of action may be effective if the herbicides have high efficacy in controlling the target weeds. Similarly, herbicides with different modes of action or methods that combine herbicides and mechanical weed control (tillage) may be used sequentially to control the same generation (i.e., emergence cohort) of weeds.

The selective differential between individuals with and without resistance alleles can also be reduced by rotating the types of herbicides used to control the target weeds so that selection for resistance to a specific herbicide occurs only in alternate growing seasons (Jasieniuk et al., 1996; Roux et al., 2008). When no fitness costs[8] are associated with resistance, the rotation of herbicides contributes to equalizing the fitness of individuals that are resistant to and susceptible to a herbicide during seasons when the herbicide is not used. Models suggest that the evolution of resistance to the rotated herbicides will be delayed by 1 year for each year that the rotation tactic is used (Maxwell and Jasieniuk, 2000). When fitness costs are associated with resistance to a herbicide (Gressel and Segel, 1990; Jasieniuk et al., 1996; Baucom and Mauricio, 2004), the fitness of individuals that have resistance alleles is lower than the fitness of individuals that do not during seasons when the herbicide is not used. Therefore, herbicide rotation contributes to reducing the selective differential between individuals with and without resistance alleles over time, which may delay the evolution of resistance (Jasieniuk et al., 1996; Roux et al., 2008).

Reduction in the selective differential can be accomplished by rotating the type of crops grown in a field between growing seasons; this may result in drastic changes in the types of herbicides used. Changes in ecological conditions associated with cultivation of different crops could favor declines in particular weed species (which could be resistant or tolerant to glyphosate) or induce competitive disadvantages in herbicide-resistant weeds through negative cross-resistance, in which resistance to one chemical confers hypersensitivity to another chemical (Gressel and Segel, 1990; Boerboom, 1999; Owen and Zelaya, 2005; Beckie et al., 2006; Murphy and Lemerle, 2006).

---

[8]Some genes that confer resistance affect the biological or physiological viability of an organism adversely in absence of a pesticide, and carriers of such a gene tend to become rarer in the population over time. The fitness cost is the extent to which such a penalty on fitness exists.

Developing Weed-Management Strategies for Herbicide-Resistant Crops

How might the various strategies be used in the context of HR cropping systems? Tank-mixes and sequences of herbicides to extend the useful life of herbicides could be employed if crop cultivars that are resistant to two or more herbicides are developed; this strategy is currently favored by biotechnology companies (Duke, 2005; Behrens et al., 2007; Green et al., 2008; Green, 2009). As for using crop rotations, the increasingly common practice of farmers throughout the United States of using glyphosate as the primary or only weed-management tactic in rotations of different glyphosate-resistant crops limits the application of the rotation strategy, even if the change in crop-induced ecological changes might improve weed management. A possible solution could be to combine the rotation of two or more HR cultivars that each can tolerate only one herbicide with the use of a different herbicide at each rotation. For example, different varieties of GE canola (*Brassica napus* L.) grown in the prairie provinces of Canada were engineered for resistance to glufosinate or glyphosate. That allowed producers to include two types of HR canola into a canola–wheat–barley rotation so that canola resistant to glufosinate or glyphosate would be grown only once every 4 years in a particular field (Powles, 2008). In contrast with corn, soybean, and most cotton production, growing crop species like canola, in which hybridization between the crop and weedy relatives is possible, poses a risk of gene flow between the HR crop and the weedy relatives (Beckie et al., 2003; Légère, 2005; see also "Gene Flow Between Genetically Engineered Crops and Related Weed Species").

The same rotation strategy could be used with HR crops that are resistant to two (or more) herbicides; the same crop would be grown twice during the rotation cycle, but each of the two herbicides that it can resist would be applied only every other year. Genes that confer resistance to some acetyl-CoA carboxylase (ACCase) inhibitors, synthetic auxins (e.g., 2, 4-D), acetolactate synthase (ALS) inhibitors, dicamba, glufosinate, glyphosate, and hydroxyphenylpyruvate dioxygenase (HPPD) inhibitors are the most likely candidates for production of the next generation of HR varieties that are resistant to multiple herbicides (Duke, 2005; Behrens et al., 2007; Green, 2009). So far, weed resistance to glufosinate and HPPD inhibitors has not been reported. Weed resistance to dicamba has not been reported in corn, cotton, or soybean but has appeared in other crops in the United States (Heap, 2010). However, weed resistance to some ACCase inhibitors, synthetic auxins, and ALS inhibitors has been reported in corn, cotton, and soybean (Heap, 2010). Moreover, most weed species that have evolved resistance to glyphosate in fields of HR crops (Table 2-1) also have evolved resistance to ALS inhibitors (Heap, 2010).

From the point of view of herbicide-resistance management and the long-term efficacy of an HR crop, it may be better to engineer a crop for resistance to herbicides that can efficiently control most weeds associated with the crop. For example, genes that confer resistance to ALS inhibitors, to which many weed species are already resistant, could be inferior to genes that confer resistance to dicamba, glufosinate, and HPPD inhibitors to produce durable HR corn, cotton, and soybean resistant to two or more herbicides. Similarly, care should be taken to engineer crops for resistance to specific ACCase inhibitors and synthetic auxins that will still be effective in controlling weeds associated with future HR crops.

If crops that are resistant to multiple herbicides—including ALS inhibitors, ACCase inhibitors, synthetic auxins, and glyphosate—are widely planted, continued use of the herbicides in fields that contain weeds already resistant to some of them could involve a risk of selecting for high levels of multiple herbicide resistance. The ability of weeds to evolve biotypes that have multiple herbicide resistance has already been demonstrated in waterhemp populations in Illinois and Missouri that are resistant to three herbicide mechanisms of action (Patzoldt et al., 2005; Legleiter and Bradley, 2008). Evolved multiple resistance will exacerbate problems of controlling some key herbicide-resistant weeds, and local and regional spatially explicit information on the distribution of weeds that are resistant to glyphosate and other herbicides could be useful in helping to manage such a situation (Werth et al., 2008). Tank-mixes and sequencing herbicides rely on redundancy to be effective. Models assessing sequential use of herbicides only, or of herbicides and mechanical weed control, indicate that a low frequency of alleles conferring resistance to herbicides and high weed mortality are critical factors for these strategies to substantially delay the evolution of weed resistant to glyphosate in HR crops (Neve et al., 2003; Neve, 2008; Werth et al., 2008).

In conclusion, regardless of the specific herbicide for which HR crops are genetically engineered, only appropriate stewardship by the grower will delay the evolution of resistance to the herbicide. Resistance management is voluntary in the United States for all pesticides except Bt produced by Bt crops (Berwald et al., 2006; Thompson et al., 2008). Given the rapid increase in and expansion of weeds that are resistant to glyphosate in HR crops, herbicide-resistance management needs national attention. As discussed previously, the rapid evolution of weed resistance to glyphosate has probably been a consequence of growers' management decisions that favored the use of glyphosate as the primary, if not sole, tactic to control weeds despite efforts in the private and public sectors to strongly recommend alternative strategies (Johnson et al., 2009). Without changes in production practices, the increase in weeds resistant to glyphosate will likely increase weed-management expenses for farmers. The evolution of herbicide resistance and other weed shifts associated with the adoption of GE crops requires the development and use of more effective weed-management strategies and tactics (Beckie, 2006; Murphy and Lemerle, 2006; Green et al., 2008; Gustafson, 2008; Powles, 2008; Werth et al., 2008).

Diversification of weed-management strategies can be accomplished by integrating several weed-control tactics: herbicide rotation, herbicide application sequences, and the use of tank-mixes of more than one herbicide; the use of herbicides that have different modes of action, methods of application, and persistence; cultural and mechanical control practices; and equipment-cleaning and harvesting practices that minimize the dispersal of herbicide-resistant weeds. Although the strategies to mitigate weed shifts are readily identified, they have largely been ignored because of the scale of commercial agriculture, which favors the simplicity, convenience, and short-term success of herbicide use over more time-consuming strategies that can be burdensome to implement on farms (Shaner, 2000; Mueller et al., 2005; Johnson and Gibson, 2006; Sammons et al., 2007; Owen, 2008). Furthermore, increased reliance on glyphosate for weed control in glyphosate-resistant crops has reduced the price of other herbicides in the United States and has limited efforts to develop new herbicides (Shaner, 2000; Duke, 2005). Companies are increasingly focused on expanding the use of currently registered herbicides, which can be achieved by commercializing GE crops that are resistant to more than one herbicide (Duke, 2005; Green, 2007, 2009). Delaying the evolution of resistance to herbicides that are used with HR crops and minimizing other weed shifts are particularly important in this context because new herbicides may not be readily available to replace ones that become ineffective when resistance evolves. Therefore, farmers would benefit from focusing on more diverse, longer-term weed-management strategies to preserve the effectiveness of HR crops and to minimize the possibility of more expensive control tactics in the future.

ENVIRONMENTAL IMPACTS OF INSECT-RESISTANT CROPS

The adoption of Bt crops has changed insect-management strategies for most corn and cotton farmers in the United States. Those changes have implications for pest populations, soil conditions, and the management of insect pests in the future. The following section evaluates the impact of insect-resistant (IR) crop adoption on pest populations, on nontarget insects, and on soil quality. It also investigates resistance-management strategies and concerns related to the continued effectiveness of IR crops.

Levels of Insecticide Use

Insecticide Use in Corn

Insecticide use in corn (in pounds of active ingredient per acre) has steadily declined since 1997 as the adoption of Bt corn (which reached 50 percent of corn acres planted in 2007) has increased (Figure 2-7). Bt corn was introduced in the mid-1990s to control European corn borer (*Ostrinia nubilalis*). Because chemical control of European corn borer was not always profitable (and timely application was difficult) before the introduction of Bt corn, many farmers accepted yield losses rather than incur the expense and uncertainty of chemical control. For those farmers, the introduction of Bt corn resulted in yield gains rather than pesticide savings (Fernandez-Cornejo and Caswell, 2006). However, a new type of Bt corn introduced in 2003 to protect against corn rootworm (*Diabrotica* spp.), which was previously controlled with chemical insecticides and crop rotation, has provided substantial insecticide savings (Fernandez-Cornejo and Caswell, 2006).

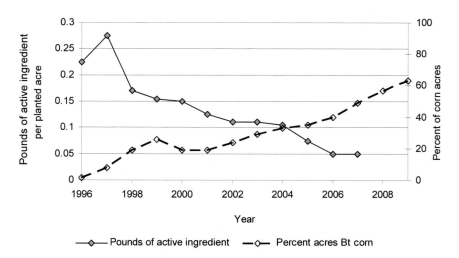

FIGURE 2-7 Pounds of active ingredient of insecticide applied per planted acre and percentage of acres of Bt corn, respectively.
NOTE: Seed-applied insecticide not included. Furthermore, the strong correlation between the rising percentage of Bt corn acres planted over time and the decrease in pounds of active ingredient per planted acre suggests but does not confirm causation between these variables.
SOURCE: USDA-NASS, 2001, 2003, 2005, 2007, 2009a, 2009b; Fernandez-Cornejo et al., 2009.

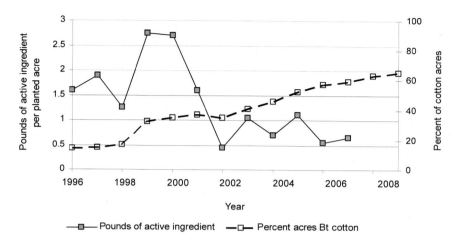

FIGURE 2-8 Pounds of active ingredient of insecticide applied per planted acre and percentage of acres of Bt cotton, respectively.
NOTE: The strong correlation between the rising percentage of Bt cotton acres planted over time and the decrease in pounds of active ingredient per planted acre suggests but does not confirm causation between these variables.
SOURCE: USDA-NASS, 2001, 2003, 2005, 2007, 2009a, 2009b; Fernandez-Cornejo et al., 2009.

Insecticide Use in Cotton

Cotton has the highest traditional use of insecticides per acre and the highest rate of adoption of Bt crops, reaching almost 60 percent in 2007, 12 years after Bt cotton was first commercialized (Figure 2-8). Insecticide use has fallen (in pounds of active ingredient per acre) over the same period, but fluctuations in total cotton insecticide applications have also been strongly affected by the boll weevil eradication program[9] (Fernandez-Cornejo et al., 2009).

---

[9]Since the 1970s, cotton growers and governments have worked toward eradicating the boll weevil, a beetle that affects cotton and that is not directly affected by Bt cotton. Different cotton-growing regions joined the program in different years. Typically, the first year of participation entails heavy application of pesticides (generally malathion). In subsequent years, the boll weevil population is monitored and treated as needed. A new wave of cotton-growing regions began participating in 1993. The spike in cotton insecticide applications in 1999 and 2000 coincides with the entry of 2 million cotton acres into the program in Texas (Fernandez-Cornejo et al., 2009).

Regional Pest Reductions

Corn and cotton that produce Bt toxins can cause high mortality in insect pest populations in which Bt-resistance alleles are rare. For example, mortality in pink bollworm (*Pectinophora gossypiella*) and tobacco budworm (*Heliothis virescens*) on Bt cotton with Cry1Ac protein is virtually 100 percent throughout the growing season (Tabashnik et al., 2000; Showalter et al., 2009). However, mortality in the moths *Helicoverpa armigera* and *Helicoverpa zea*[10] on Cry1Ac cotton is typically lower than 95 percent and declines during the growing season (Kennedy and Storer, 2000; Olsen et al., 2005; Tabashnik et al., 2008; Showalter et al., 2009). Crops that produce more than one Bt toxin generally cause higher mortality than crops that produce a single toxin although declines in mortality during the growing season may still be observed (Adamczyk et al., 2001; Bommireddy and Leonard, 2008; Mahon and Olsen, 2009; Showalter et al., 2009).

Because Bt crops can cause high pest mortality, it has been postulated that one effect of the widespread use of Bt crops is a reduction in some pest populations regionally (Kennedy et al., 1987; Alstad and Andow, 1995; Roush, 1997; Gould, 1998; Kennedy and Storer, 2000; Storer et al., 2003). According to that idea, an area-wide decline in pest abundance could occur because replacing non-Bt crop fields with Bt crop fields eliminates suitable habitats for the pests. If females lay eggs on Bt plants and on non-Bt host plants, laying eggs on Bt plants could substantially reduce the number of surviving offspring produced by females and cause a decline in pest density (Riggin-Bucci and Gould, 1997; Carrière et al., 2003; Shelton et al., 2008). Models have suggested that the suppression of pest populations is more likely as mortality induced by Bt crops increases, the abundance of Bt crops and female movement between patches of Bt and non-Bt plants increase, and the net reproductive rate in patches of non-Bt hosts decreases (Carrière et al., 2003). However, polyphagous pest species (those able to feed on multiple types of plants) often exploit crops sequentially during the growing season, tracking changes in host suitability (Kennedy and Storer, 2000). In some cropping systems, the feeding options of such pests might be limited to only a Bt crop for a few generations, when it is the only suitable resource available. Thus, Bt crops could affect pest population dynamics even when the crops are relatively rare (Kennedy et al., 1987; Wu et al., 2008).

Long-term monitoring of insect-pest density before and after commer-

---

[10]*Helicoverpa armigera* and *Helicoverpa zea* are known by many common names, depending on the host plant. *Helicoverpa armigera* is referred to as old world bollworm or cotton bollworm (when it feeds on cotton), pod borer (when it feeds on chickpea or pigeon pea), tomato fruit borer (when it feeds on tomato), and corn earworm (when it feeds on corn). *Helicoverpa zea* is often called cotton bollworm, corn earworm, or tomato fruitworm.

cialization of Bt crops has provided evidence that deployment of Bt crops influences pest population dynamics regionally. Table 2-3 contains the results of pest-monitoring studies in the United States and China. Most of the studies covered a single region where spatially explicit data on the distribution of Bt crops were not available, but in one study of pink bollworm population density in Arizona from 5 years before to 5 years after introduction of Bt cotton, the abundance of Bt and non-Bt cotton fields in 15 cotton-growing regions was quantified with geographical information system technology (Carrière et al., 2003). In regions with less than an average of 65 percent Bt cotton in the second 5-year period, the introduction of Bt cotton had no consistent effect on population density of pink bollworm; in regions with more than 65 percent Bt cotton, the introduction of Bt cotton decreased pink bollworm population density, and the extent of the decline increased as the percentage of Bt cotton increased. Those data are consistent with modeling results and suggest that pest-population suppression occurs if the area of Bt crops exceeds a threshold percentage of Bt cotton (Carrière et al., 2003). Another recent study of European corn borer conducted in five major U.S. corn-producing states indicated that suppressive effects of Bt corn depended on the extent of adoption of the technology (Hutchison et al., 2007).

Storer et al. (2008) noted that producers, extension agents, and pesticide appliers reported less serious insect-pest control problems in non-Bt crops, such as soybean and vegetables, after the regional suppression of European corn borer and corn earworm (*H. zea*) by Bt corn in Maryland. As a comparison to the U.S. experience, a study conducted in six provinces of China from 1997 to 2006 documented a progressive decline in the population density of cotton bollworm (*H. armigera*) after the introduction of Bt cotton (Wu et al., 2008; Table 2-3). The significant suppression of cotton bollworm occurred not only in Bt and non-Bt cotton but in corn, peanut, soybean, and vegetables. Wu et al. (2008) proposed that the regional decline of cotton bollworm populations could reduce insecticide use in crops other than cotton. Nevertheless, the economic consequences of the regional suppression of pests by Bt crops have been investigated only for European corn borer in five U.S. Corn Belt states (Hutchison et al., 2007). It was estimated that regional declines in European corn borer population densities during the last 14 years in those states saved at least $3.9 billion for producers of non-Bt corn and $6.1 billion for producers of Bt and non-Bt corn combined. Further detailed spatially explicit studies of the association between the distribution of Bt crops and pest problems on different scales will be helpful in improving understanding of how the use of Bt crops can reduce pest abundances (Marvier et al., 2008).

The use of Bt crops sometimes changes pest-management practices enough to increase problems related to pests that are not killed by Bt

TABLE 2-3 Regional Effects of Deployment of Bt Crops on Population Dynamics of Major Pests of Corn and Cotton

Pest	Location	Regional Use of Bt Crops	Number of Years Monitored	Population Decline	Reference
Heliothis virescens	Mississippi (Washington County)	Up to 85% Bt cotton	19 (1986–2005)	Yes	Adamczyk Jr. et al., 2001
Helicoverpa zea	Mississippi (Washington County)	Up to 85% Bt cotton	19 (1986–2005)	Yes	Adamczyk Jr. et al., 2001
Heliothis virescens	Louisiana (Bossier City)	Unknown	20 (1986–2005)	Yes	Micinski et al., 2008
Helicoverpa zea	Louisiana (Bossier City)	Unknown	20 (1986–2005)	No	Micinski et al., 2008
Ostrinia nubilalis	Maryland (Eastern Shore)	Up to 60% Bt corn	35 (1973–2007)	Yes	Storer et al., 2008
Helicoverpa zea	Maryland (Eastern Shore)	Up to 60% Bt corn	35 (1973–2007)	Yes	Storer et al., 2008
Helicoverpa armigera	China (Six provinces)	Up to 10% Bt cotton	10 (1997–2006)	Yes[a]	Wu et al., 2008
Ostrinia nubilalis	Minnesota, Illinois, Iowa, Nebraska, Wisconsin	Up to 75% Bt corn	ca. 45 years (1943–2007)[b]	Yes	Hutchison et al., 2007
Pectinophora gossypiella	Arizona (15 regions)	0.01–87% Bt cotton	10 (1992–2001)	Yes	Carrière et al., 2003

[a]Study analyzed decline in density of *H. armigera* after commercialization of Bt cotton.
[b]Data analyzed for about 45 years in each state. Not all states were monitored each year in the period 1943–2007.

toxins. For example, substantial reductions in the use of synthetic insecticides on Bt cotton favored outbreaks of mirids and leafhoppers in China (Wu et al., 2002; Men et al., 2005). Those pests had been well controlled by insecticides before the introduction of Bt cotton. Similarly, lower use of insecticides in Bt cotton probably contributed to the higher stink bug damage in cotton in some southern U.S. states although the regional increases in stink bug populations were probably influenced by other factors as well (Greene et al., 2001, 2006). Changes in pest-management practices in connection with Bt crops can also have favorable consequences for the control of some pests that are not killed by Bt toxins. For example, a reduction in insecticide use on Bt cotton was sometimes associated with greater predator abundance and better pest control in cotton aphid in the United States (see section "Natural Enemies").

Reversal of Insect Resistance to Synthetic Insecticides

The deployment of Bt crops is known to promote a reversal of pest resistance to synthetic insecticides, but this has not yet been observed in the United States. In northern China, the reduction in use of insecticides on Bt cotton contributed to restoring cotton bollworm (*H. armigera*) susceptibility to some synthetic insecticides (Wu et al., 2005; Wu, 2007) although fitness costs associated with insecticide resistance likely helped to increase susceptibility. Similarly, resistance to pyrethroid insecticides declined considerably in tobacco budworm (*H. virescens*) after the introduction of Bt cotton in southern Tamaulipas, Mexico (Terán-Vargas et al., 2005). The renewed efficacy of insecticides provided more pest-management options to producers in those regions. However, such reversals in insecticide resistance do not always occur. For example, the planting of Bt cotton in Louisiana did not change the high levels of pyrethroid resistance in tobacco budworm (*H. virescens*) and cotton bollworm (*H. zea*) (Bagwell et al., 2001), and *H. zea* resistance to pyrethroids increased substantially after the planting of Bt cotton in several regions of Texas (Pietrantonio et al., 2007).

Effects on Nontarget Species

Bt toxins are considered acutely toxic to a relatively narrow array of invertebrate taxa when compared with broad-spectrum insecticides because toxicity through direct ingestion of a Bt toxin is typically restricted to insects in the same order as the target pest (Schnepf et al., 1998; Glare and O'Callaghan, 2000; van Frankenhuyzen and Nystrom, 2002; Mendelsohn et al., 2003). For example, the endotoxins Cry1Aa, Cry1Ab, and Cry1Ac kill mainly particular moths and butterfly species, while Cry3Aa and

Cry3Bb mainly kill particular beetle species. Furthermore, because Bt toxins are specific, they cause different mortality within targeted insect orders. For example, the cotton cultivar Bollgard I®, which produces the toxin Cry1Ac and targets lepidopteran pests, kills virtually 100 percent of pink bollworm and tobacco budworm (*H. virescens*), between 24–95 percent of cotton bollworms *H. zea* and *H. armigera*, and less than 4 percent of fall armyworm (*Spodoptera frugiperda*) and beet armyworm (*Spodoptera exigua*) (Showalter et al., 2009). Field studies have revealed relatively few adverse effects of Bt crops on arthropods that are not closely related to the target pests (Cattaneo et al., 2006; Romeis et al., 2006). In contrast, broad-spectrum insecticides, such as pyrethroids and organophosphates had consistent, adverse effects on a wide array of nontarget arthropods (Cattaneo et al., 2006; Romeis et al., 2006).

Although the high specificity of Bt crops for the control of target pests is consistent with integrated pest management, they may have effects on beneficial organisms. For example, the larvae of nontarget moths or butterflies in the landscape surrounding farms may be susceptible to Bt toxins that target pests in this group, but they would need to eat the Bt plant material to be affected. Bt corn byproducts that enter streams may affect aquatic insects in related taxa (Rosi-Marshall et al., 2007). The abundance of some natural enemies may decrease when their host or prey species are susceptible to Bt toxins and as a result become rare or nutritionally less suitable (Romeis et al., 2006).

Quantifying and predicting the effects of Bt crops on nontarget invertebrate species has been the subject of considerable work. As compiled by Marvier et al. (2007) and Naranjo (2009), research on the nontarget effects of Bt crops includes 135 laboratory studies of nine Bt crops and 22 Bt Cry proteins or protein combinations and 63 field studies of five Bt crops and 13 Bt proteins. In total, field and laboratory studies of at least 99 and 185 invertebrate species, respectively, have been conducted although not with equal effort. Most of the field studies have been of corn and cotton. Individual study results vary, so evidence-based generalizations are elusive in the absence of formal approaches. A review of recent syntheses provides an overview of the generalizations that have emerged thus far from those efforts.

For cotton and corn, whether Bt crop fields have more or fewer nontarget invertebrates depends on whether one compares the Bt crop to a conventional counterpart that received insecticide treatments (Marvier et al., 2007; Wolfenbarger et al., 2008; Naranjo, 2009). Collectively, studies have indicated that a higher total abundance of arthropods occurred in Bt fields than in conventional fields sprayed with insecticides and a lower abundance than in conventional fields with no insecticide treatment (Marvier et al., 2007). For Bt corn, the magnitude of the effect also depended on whether studies tested Bt176 (no longer registered for use) or the MON810 (commercially used) Bt events. Lower abundance of specific taxa was found in Bt fields than in unsprayed, non-Bt fields; the taxa in question included moths, butterflies, beetles, and true bugs on cotton and wasps on corn. Differences in the availability of prey or in survival may explain those results (Marvier et al., 2007).

Bt potato crop fields without insecticide use contained higher abundances of predators, natural enemies as a whole, and nontarget pests compared to conventional potato fields, whether or not insecticides were applied to the conventional fields (Wolfenbarger et al., 2008).

Natural Enemies

Maintenance of healthy populations of predators of crop pests is a desirable goal for ensuring long-term environmental sustainability of farms. Decreasing the numbers of predators, which in practice will be related to the overall biodiversity in an area and to on-farm pest control (Landis et al., 2008), would be undesirable. Even in systems where a single predator may suffice as a biocontrol agent, redundancy is an important tool for ensuring ecosystem services.

The few studies comparing biological control (by parasitism and predation rates) between Bt and conventional crops have suggested that control of nontarget pests on Bt crops was enhanced on cotton (Head et al., 2005) or similar on cotton (Naranjo, 2005) and corn (Pons and Starý, 2003; Naranjo, 2005) and that control of target pests on Bt crops was enhanced on cotton (Head et al., 2005), similar on cotton (Sisterson et al., 2004b; Naranjo, 2005) and corn (Orr and Landis, 1997; Sisterson et al., 2004b; Naranjo, 2005), or reduced on corn (Siegfried et al., 2001; Bourguet et al., 2002; Manachini, 2003; cited by Naranjo, 2009). Maintenance of biological control of nontarget pests in one study occurred in Bt cotton fields in spite of about a 20-percent reduction in the abundance of some common predators (Naranjo, 2005). When Bt crops have completely replaced insecticide-treated conventional crops, studies have consistently reported higher numbers of predators on cotton, corn, and potato. When Bt crops have replaced non–insecticide-treated conventional crops, results studies have consistently indicated slightly fewer predators on Bt cotton and no detectable difference on Bt corn (Wolfenbarger et al., 2008; Naranjo, 2009).

Field studies of parasitoids have overemphasized specialist species of the target pest of Bt corn, so generalizations to parasitoids as a group are premature. The studies have revealed a pattern similar to that of predators: fewer parasitoids in conventional corn fields sprayed with insecticides and no detectable difference between Bt corn fields and conventional corn fields not treated with insecticides. Laboratory studies indicate that effects of Bt toxins on parasitoids depend on whether they are fed prey that are susceptible to Bt toxins (Zwahlen et al., 2000; Dutton et al., 2002; Schuler et al., 2003, 2004; Romeis et al., 2006). Syntheses of laboratory studies of 14 parasitoid species indicate a favorable or neutral effect on life-history traits when they were fed prey that had ingested a Bt toxin but were not affected by it (high-quality prey). Conversely, studies have shown longer development times, lower reproduction, and lower survival if the parasitoids were fed prey that had ingested a Bt toxin that was toxic to them (low-quality prey) (Naranjo, 2009).

The adoption of Bt cotton increases abundances of natural enemies and hence the potential for biological control when it completely replaces insecticide treatments. Moth larvae were responsible for a large fraction of cotton-insect losses before the adoption of Bt cotton, but cotton-insect losses caused by these larvae have become less important now that Bt cotton has been widely adopted. The five major insect pests of cotton in the United States in 2008 were lygus bugs (1 percent yield loss), boll-worms and budworms (0.76 percent yield loss), stink bugs (0.75 per-cent yield loss), thrips (0.52 percent yield loss), and cotton fleahoppers (0.23 percent yield loss) (Williams, 2009). Among those, only bollworms and budworms are controlled by Bt cotton, so the use of Bt cotton rarely eliminates all insecticide applications. Actual farm-level reductions in insecticide use for Bt cotton would probably increase the abundance of nontarget insects less consistently (e.g., Cattaneo et al., 2006; Sisterson et al., 2007) than what has been observed in experimental studies in which the use of Bt cotton completely replaced insecticide treatments.

Pollinators and Other Valued Insects

The honey bee (*Apis mellifera*) is one of the agricultural sector's most important pollinators. Laboratory toxicity studies of honey bees have consistently found no evidence that Bt pollen or Bt proteins decrease honey-bee larval or adult survival (Duan et al., 2008) even at toxin con-centrations well beyond what would be encountered in the field. There have been laboratory or field studies of few other species (Wolfenbarger et al., 2008; Naranjo, 2009); no consistent effect on development time (eight studies) or survival (20 studies) has been detected in labora-tory tests, but effects varied widely among studies, particularly for development time (Naranjo, 2009). Laboratory studies collectively have indicated longer development time and lower survival of valued insect herbivores, a category that includes charismatic species (e.g., monarch butterfly larvae) and moths of economic importance (e.g., the silkworm) (Naranjo, 2009).

Summary of Nontarget Effects

The abundance of natural enemies on Bt crops can be greater than, the same, or lower than on non-Bt crops. The magnitude of the benefit depends on the extent to which a Bt crop substitutes for the use of insecticide treatments of non-Bt crops and on whether insecticides for other pests are used on the Bt crop. Honey-bee adults and larvae were not harmed by Bt pollen or Bt proteins, but too few pollinators have been studied to support generalizations about the group as a whole. As the sophistication of GE-crop varieties increases and the functional roles of arthropods become understood more fully, it should be possible to develop strategic pest-management systems that maintain high crop productivity while avoiding effects on nontarget moths, butterflies, and beetles.

Soil Quality

Overall, it appears that current Bt crops have no greater or lesser effect on soil quality than the crops that they have replaced. Many peer-reviewed studies have addressed the nontarget impacts of Bt crops on soil organisms. Specifically, studies have considered the effects of plant residues on the soil community because plants are the primary source of carbon in soils. If Bt toxins affect soil microorganisms, rates of decomposition and nutrient cycling may be altered. Studies have also focused on the consequences of Bt-containing root exudate. Root exudate influences the soil community, especially the community of distinct, specialized soil microorganisms associated with roots.

Most assessments of the effects of Bt insecticidal proteins on soil microorganisms and other organisms have found that these proteins do not substantially alter microbial populations and measured functions (Icoz and Stotzky, 2008). Over four years of continuous corn cultivation, Bt plant residues and root exudates had no consistent or persistent effect on a breadth of microorganisms or their enzymatic activity in the soil, but differences were detected according to plant species, variety, and age as well as other environmental factors (Icoz et al., 2008). With respect to macro-organisms, Lang et al. (2006) found no significant differences in earthworm or springtail population density or biomass between soils with Bt and with non-Bt corn or between soils with corn treated and not treated with insecticide (baythroids) at five sites during 4 years of corn cultivation. Instead, the site and the sampling years had a greater influence on earthworm population density and biomass than the presence of the Cry protein. Those results corroborate other laboratory and field studies of the effect of Bt toxins on survival, growth, and reproduction of an array of soil invertebrates, including woodlice, springtails, and mites (Ahl Goy et al.,

1995; Saxena and Stotzky, 2001a; Zwahlen et al., 2003; Clark et al., 2006; Vercesi et al., 2006; Krogh et al., 2007). Similarly, Birch et al. (2007) detected transient and site-specific reductions in the biomass of oribatid mites and total microarthropods in fields under Bt and non-Bt corn. However, the differences between populations under different non-Bt corn varieties were often of the same magnitude as those between Bt and non-Bt corn; this led to the conclusion that the effects in the field were varietal effects and not due specifically to the Bt trait (Cortet et al., 2006).

When the effects of Bt toxins on nematodes were studied, season, soil tillage, soil type, crop type, and cultivar influenced nematode number to a greater extent than whether the corn was a Bt variety. Under field cultivation for Bt crops with the Cry3Bb1 protein and non-Bt crops, no effect was detected on the abundance of the nematode *Caenorhabditis elegans* on corn (Al-Deeb et al., 2003) or on the relative abundance of species of nematodes in the soil with eggplant (Manachini, 2003; cited by Icoz and Stotzky, 2008). When soils from Bt corn with the Cry1Ab protein and non-Bt corn in cultivation were compared, there were no effects on nematode communities and diversity (Manachini, 2003; cited by Icoz and Stotzky, 2008) or on the nematode *Pratylenchus* spp. (Lang et al., 2006). However, in experiments in cultivated Bt and non-Bt corn fields, adverse effects on growth and abundance of *C. elegans* were observed (Manachini, 2003; Lang et al., 2006; cited by Icoz and Stotzky, 2008), and a lower abundance of natural populations occurred transiently in Bt fields and consistently at one Bt site (Griffiths et al., 2005, 2006).

Similarly, the Cry1Ab insecticidal protein for European corn borer control had less effect on the bacterial community structure than other environmental factors (Baumgarte and Tebbe, 2005). In one study, a transient decrease occurred in the numbers of protozoa in soil with Bt corn under field conditions (Griffiths et al., 2005); otherwise, no toxic effects of the Cry proteins on protozoa have been observed (Donegan et al., 1995; Saxena and Stotzky, 2001a; Griffiths et al., 2005; Icoz and Stotzky, 2008). No changes in microbial activity and other assays (i.e., nitrogen mineralization potential, short-term nitrification, and soil respiration rate) occurred when soils cropped with corn that produced the Cry3Bb toxin for corn rootworm protection were compared with soils cropped with the nontransgenic isoline (Devare et al., 2004, 2007). Those studies have indicated that Bt and non-Bt crops have comparable effects on soil bacteria and protozoa.

Rates of residue decomposition and the associated accumulation of soil organic matter affect soil productivity and soil ecological functions; therefore, if residues of Bt corn differ from residues of non-Bt corn in decomposition rates, there might be long-term implications for soil quality and soil carbon sequestration. Reduced decomposition rates might increase the time that Bt toxins remain in the environment. Chemical bonds in lignins are more resistant to microbial decomposition than other chemical bonds in plant cells. Some studies have demonstrated higher lignin content in Bt corn hybrids compared to their respective isolines (or near-isolines) (Saxena and Stotzky, 2001b; Poerschmann et al., 2005), but other studies have found no differences (Jung and Sheaffer, 2004). Decomposition rates have more importance for soil quality than relative lignin contact. In some laboratory research, plant residue of Bt hybrids decomposed at a lower rate in soil than residue of non-Bt hybrids (Flores et al., 2005), but field studies have not detected differences in decomposition rates (Lehman et al., 2008). Similarly, Hopkins and Gregorich (2003) reported no differences in carbon dioxide production from Bt and non-Bt corn in soil over a 43-day incubation period. No differences in mass losses in the field were detected between Bt-glyphosate–resistant and glyphosate-resistant cotton lines, indicating that there were no differences in the rate of decomposition or change in nutrient content in the litter over the 20-week experiment (Lachnicht et al., 2004). When the whole soil organism community was allowed to access the residue, the decomposition of Bt and non-Bt residue was similar (Zwahlen et al., 2007). Tarkalson et al. (2008) also reported no differences in residue decomposition rates or in mass of total carbon remaining over time between Bt and non-Bt corn hybrids observed in a field study although the study did detect differences in rates of decomposition for leaf, stalk, and cob plant parts. Finally, after seven years of continuous corn cultivation, no differences between Bt and non-Bt corn treatments were detected in total carbon or nitrogen in soil, indicating that plant decomposition rates were similar (Kravchenko et al., 2009). On the basis of those studies, the plant residue from Bt and non-Bt corn hybrids decomposed at similar rates and would have similar effects on soil quality and on potential carbon sequestration.

Evolution and Management of Insect Resistance

Evolution of Resistance

Insects can adapt to toxins and other tactics used to control them (Palumbi, 2001; Onstad, 2008). When Bt crops were first considered for commercial introduction, EPA recognized their potential to reduce human and environmental exposure to broad-spectrum insecticides, increase growers' ability to manage pests and improve crop quality, and increase profits at the farm and industry levels (Berwald et al., 2006; Matten et al., 2008). Those benefits had already been demonstrated by sprayed Bt insecticides that are critical pest-management tools for many fruit and vegetable crops in the United States (Walker et al., 2003a). As the regulatory agency overseeing the introduction of biological pesticides, EPA concluded that the potential for rapid evolution of insect resistance to Bt toxins produced by GE crops was a threat to the benefits provided by Bt crops and to the efficacy of Bt sprays in organic and conventional production systems (Matten et al., 2008; Thompson et al., 2008). Accordingly, it mandated the use of a refuge strategy (described later in this chapter) to delay the evolution of resistance in major insect pests controlled by Bt corn and cotton (US-EPA, 2008a).

Extensive monitoring of 11 major lepidopteran pests of corn and cotton over the last 14 years has revealed that some populations of one moth species, cotton bollworm (*Helicoverpa zea*), evolved resistance to the Bt toxins Cry1Ac and Cry2Ab found in some cotton cultivars in the United States (Tabashnik and Carrière, 2008; Tabashnik et al., 2008, 2009a). In addition, some populations of fall armyworm evolved resistance to Cry1F corn in Puerto Rico (Matten et al., 2008), and some populations of corn stem borer (*Busseola fusca*) evolved resistance to Cry1Ab corn in South Africa (van Rensburg, 2007; Kruger et al., 2009).

That resistance has evolved in only three pest species in the last 14 years suggests that the refuge strategy has successfully delayed the evolution of resistance to Bt toxins (Tabashnik et al., 2008, 2009a). Comparisons between pests that have and have not evolved resistance to Bt crops suggest that recessive inheritance of resistance and abundant refuges of non-Bt host plants are two key factors that delay the evolution of resistance (Tabashnik et al., 2008, 2009a). In accordance with these findings, EPA demands that GE seed companies require producers to plant refuges to delay the evolution of resistance where such refuges are deemed necessary and develop compliance assurance programs (Thompson et al., 2008). The promotion of precise resistance-management guidelines by the industry has undoubtedly contributed to increasing the use of refuges for managing the evolution of resistance to Bt crops in the United States. In some regions, compliance to the mandated refuge strategies has been high since the introduction of Bt crops (Carrière et al., 2005). However, levels of compliance have substantially and regularly declined in other parts of the country, possibly because the use of Bt crops has increased globally and producers can no longer rely on non-Bt users to provide refuges for their farms (Jaffe, 2009).

Although the theory, resistance-monitoring data, and experimental work conducted in the laboratory suggest the refuge strategy has been useful, detailed field experiments are still needed to demonstrate how the refuge strategy can delay the evolution of resistance to Bt crops. There is usually a delay between the introduction of a novel pesticide and the rapid rise in the number of species that have evolved resistance to it (Georghiou, 1986). That is illustrated in a comparison of the cumulative number of cotton pests that evolved resistance to Bt toxins in crops and to the insecticide dichlorodiphenyltrichloroethane (DDT) after the introduction of these pest-management tools in the United States (Figure 2-9).

After commercialization of Bt cotton in 1996, its use increased rapidly in the United States. Similarly, the use of DDT in cotton increased rapidly after it became widely commercially available in 1946. For example, 90 percent of agricultural DDT applications in the United States targeted cotton pests in 1962 (Walker et al., 2003b). Similar to Bt cotton that produces high concentrations of Bt toxins over much of the growing season, DDT was applied repeatedly in cotton and retained toxicity for extended periods (US-EPA, 2000). The recessive mutations *kdr* and super-*kdr* confer recessive resistance to DDT in many agricultural pests (Davies et al., 2007; APRD, 2009); this is similar to the inheritance of resistance to Bt toxins in cotton, which is often recessive (Tabashnik et al., 2008).

With respect to the evolution of resistance, Bt cotton and DDT differ in at least two important ways. First, DDT kills a wide array of insects regardless of their feeding habits whereas Bt cotton kills only some lepidopteran pests that feed on the cotton. Second, no refuge strategy was mandated to manage the evolution of insect resistance to DDT. Those differences suggest that the evolution of DDT resistance in cotton pests should have been more rapid than the evolution of resistance to Bt cotton. However, the cumulative number of cotton pests that evolved resistance to Bt cotton and the number that evolved resistance to DDT after their introduction in the United States have been strikingly similar (Figure 2-9). That comparison indicates that it may still be too soon to claim that the refuge strategy has substantially delayed the accumulation of pests resistant to Bt. While seed companies are in a better position to commercialize more efficient Bt cultivars for delaying the evolution of resistance (see below), the possibility remains that the accumulation of resistant pests could accelerate. Thus, complacency in the implementation of resistance-management strategies is not warranted (Hurley and Mitchell, 2008; Jaffe, 2009; Tabashnik et al., 2009a).

Principles of Population Genetics Underlying the Refuge Strategy

Population-genetic models and empirical data on factors that affect the evolution of insect resistance to Bt crops have been central in the development of the refuge strategy. Models generally assume that resistance to a toxin produced by Bt crops is conferred by mutations at a single locus (gene location) (Gould, 1998; Tabashnik and Carrière, 2008). That is a reasonable assumption because resistance to the intense selection imposed by Bt crops and insecticides is likely to involve genes that have major effects (Carrière and Roff, 1995; McKenzie, 1996). Furthermore, most of the observed cases of evolved resistance to Bt crops have involved

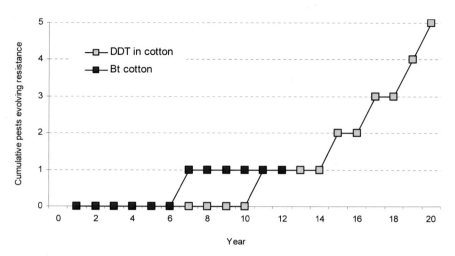

FIGURE 2-9 Cumulative number of cotton pests evolving resistance to Bt cotton and DDT in the years after these management tools became widely used in the United States.
SOURCE: APRD, 2009.

mutations at a single locus (Gahan et al., 2001; Morin et al., 2003; Yang et al., 2007; Pereira et al., 2008). For simplicity, models assume the presence of one allele that confers susceptibility and one allele that confers resistance even if more than one allele at a single locus can confer resistance to Bt crops (Morin et al., 2003; Yang et al., 2007).

The refuge strategy relies on two basic principles. The first principle is that the dominance of resistance[11] is reduced by increasing the dose of Bt toxins (Gould, 1998; Tabashnik et al., 2004). When the concentration of a Bt toxin in a plant is low, the resistance trait in the insect population is nonrecessive, but when it is high, the resistance trait in the insect population becomes recessive, and resistance becomes rarer. Accordingly, resistance to commercialized GE crops that produce high concentrations of Bt toxins is recessive in many, but not all, target pests (Tabashnik et al., 2008). The refuge strategy requires the presence of refuges of non-GE host plants in or near Bt crop fields (US-EPA, 2008a, 2008b). For refuges to be effective, the susceptible insects produced in refuges must be in sufficient

---

[11]The *dominance of resistance* depends on the response of heterozygotes compared to the response of homozygous-susceptible individuals and homozygous-resistant individuals. If heterozygotes respond like homozygous-susceptible individuals, resistance is recessive; if heterozygotes respond like homozygous-resistant individuals, resistance is dominant.

numbers and mate with the rare resistant pests that survive on GE crops. With effective refuges and recessive resistance, most hybrid offspring produced by resistant pests that survive on Bt crops are killed when they feed on GE crops. That reduces the heritability of resistance (the degree of genetic similarity between resistant parents that survive on Bt crops and their offspring) and delays its evolution (Gould, 1998; Sisterson et al., 2004a; Tabashnik and Carrière, 2008).

The second principle underlying the refuge strategy is that the evolution of resistance can be delayed or prevented by reducing the selective differential between individuals with and without resistance alleles (Gould, 1998; Carrière and Tabashnik, 2001; Andow and Ives, 2002; Tabashnik et al., 2005; Crowder and Carrière, 2009). The selective differential between resistant and susceptible individuals can be affected by crop-management practices, such as increasing refuge size that increases relative fitness of susceptible individuals (Mitchell and Onstad, 2005; Onstad, 2008). In fields of Bt crops, where resistant individuals are more abundant than susceptible individuals, the selective differential between resistant and susceptible individuals can be reduced by such crop-management practices as pheromone mating disruption and the elimination of crop residues that contain insects (Andow and Ives, 2002; Carrière et al., 2004b).

The selective differential between resistant and susceptible individuals can also be affected by pest biology and genetics. Fitness costs associated with resistance to Bt toxins occur in environments that lack Bt toxins if individuals with one or more resistance alleles have lower fitness than individuals without such alleles (Gassmann et al., 2009). Fitness costs of Bt resistance are found in many species and select against resistance in environments where Bt toxins are absent; this selection counterbalances selection that favors an increase in resistance in fields of Bt crops (Gassmann et al., 2009). Fitness costs expressed in heterozygous individuals are nonrecessive; costs expressed only in homozygous resistant individuals are recessive. Nonrecessive fitness costs can delay the evolution of resistance more effectively than recessive fitness costs because alleles that confer resistance to Bt crops are often rare (Gould et al., 1997; Andow et al., 2000; Tabashnik et al., 2006; Mahon et al., 2007), so most resistance alleles in pests targeted by Bt crops are carried by heterozygous individuals. With nonrecessive fitness costs, the fitness of resistant heterozygous individuals is lower than the fitness of susceptible individuals in refuges, and such costs can strongly select for a decline in resistance despite the fact that selection favors resistant individuals on Bt crops (Gassmann et al., 2009). In other words, recessive costs that influence only the rare homozygous resistant individuals are less effective in delaying resistance than nonrecessive costs that influence the more abundant resistant heterozygous individuals.

Incomplete resistance occurs when the fitness of resistant individuals is lower on Bt cultivars than on corresponding non-Bt cultivars (Carrière and Tabashnik, 2001). It occurs because the individuals that do survive on Bt crops are nevertheless affected adversely by Bt toxins (for example, larvae take a long time to develop on the Bt crop, and the resulting moths are smaller and less fecund). Incomplete resistance is found in many species and contributes to delaying the evolution of resistance by reducing the selective differential between resistant and susceptible individuals (Carrière and Tabashnik, 2001; Tabashnik et al., 2005; Crowder and Carrière, 2009).

The Pyramid Strategy

The first IR crops produced a single Bt toxin. More recently, the "pyramid" strategy has used GE crops that produce two distinct Bt toxins for delaying pest resistance. The pyramid strategy is based on the principle that insects are killed on two-toxin plants as long as they have a susceptibility allele at a resistance locus—a phenomenon called redundant killing (Gould, 1998; Roush, 1998). As resistance alleles are generally rare, the only genotype that has high survival on a cultivar that produces two or more Bt toxins is expected to be extremely rare. Accordingly, the refuge strategy is considered more effective in reducing the heritability of resistance when crops produce more than one Bt toxin than when they produce a single Bt toxin (Gould, 1998; Roush, 1998; Gould et al., 2006). Models suggest that the pyramid strategy is most effective when the majority of susceptible pests are killed by the GE crop, resistance to each Bt toxin is recessive, fitness costs and refuges are present, and selection with one Bt toxin does not cause cross-resistance to another (Gould, 1998; Zhao et al., 2005; Gould et al., 2006). Cross-resistance to Bt occurs when a genetically based decrease in susceptibility to one toxin decreases susceptibility to other toxins.

Changes in Refuge Strategy in the United States

In a process that aims to use scientific knowledge to balance economic and environmental considerations, refuge strategies for Bt corn and cotton mandated by EPA have been improved since the commercialization of these GE crops in 1996. EPA specifies the area, configuration, and types of refuges to be used with specific Bt crops. Changes in refuge requirements have been based on input from academe, farmers, and industry. Some of the changes have made refuge requirements more stringent, while others have eliminated refuge requirements. For example, refuge distance requirements for the use of Bt cotton against pink bollworm were unspecified from 1996 to 2000, unless the percentage of Bt cotton in a county exceeded 75 percent in the previous year (US-EPA, 1998; Carrière et al., 2001). However, on the basis of the principle that refuges must be near Bt crops to promote the desired mating between susceptible and resistant insects, new regulations enacted in 2001 limited the distance between refuges and Bt cotton regardless of the percentage of Bt cotton in the previous year (Tabashnik et al., 1999; Carrière et al., 2001; US-EPA, 2001; Carrière et al., 2004a). Since 2006, in response to a proposal from cotton growers to eradicate pink bollworm in Arizona, EPA has allowed use of mass release of sterile pink bollworm moths as an alternative to non-Bt cotton refuges (US-EPA, 2006a).

In another example, in response to a proposal from Monsanto, the refuge requirement of non-Bt cotton cultivars was abolished from Texas to the Mid-Atlantic to manage resistance of tobacco budworm (*Heliothis virescens*) and cotton bollworm (*Helicoverpa zea*) to Monsanto's pyramided Bt cotton cultivar that produces the toxins Cry1Ac and Cry2Ab (US-EPA, 2007) and subsequently to a cultivar from Dow AgroSciences producing Cry1Ac and Cry1F. The proposal included new data and modeling results that indicated that weeds and non-Bt crops other than cotton might provide sufficient refuges to delay Bt resistance in the two mobile, polyphagous pests (US-EPA, 2006b). The 2007 change by EPA meant that refuges of non-Bt cotton are no longer required for millions of acres of the Monsanto cotton cultivar grown in large areas of the United States. Assuming that no other factors changed (e.g., the technology fee), that action would improve the net benefits to farmers growing the GE cotton and increase its adoption (Luna V. et al., 2001; Matus-Cádiz et al., 2004) at least in the short term. However, it is noteworthy that decreased susceptibility to both Cry1Ac and Cry2Ab in cotton bollworm (*H. zea*) has indicated that this pest is evolving resistance to cotton producing those toxins in the United States (Tabashnik et al., 2008, 2009b).

It appears likely that most Bt crops commercialized in the future by biotechnology companies will produce two or more Bt toxins for the control of individual insect pest species (Bravo and Soberón, 2008; Matten et al., 2008). That could improve the durability of Bt crops if few other major insect pests targeted by Bt crops evolve resistance before the replacement of one-toxin crops by pyramids and if populations of pests that are resistant to single-toxin Bt crops remain rare. EPA promotes the replacement of one-toxin Bt cultivars with two-toxin Bt cultivars on the basis of recognition that the evolution of resistance is more effectively delayed with a pyramid strategy than with one-toxin crops (Matten et al., 2008). Another incentive to eliminating the use of one-toxin Bt cultivars when two-toxin Bt cultivars are introduced is that results from simulation models and small-scale laboratory experiments indicate that the evolution of resistance to two-toxin cultivars is accelerated when plants that produce two Bt toxins are grown near plants that produce just one toxin (Roush, 1998; Gould, 2003; Zhao et al., 2005).

So far, the complete replacement of one-toxin with two-toxin Bt cultivars has occurred only in Australia (Baker et al., 2008). The Transgenic and Insecticide Management Strategy committee, which comprises growers, consultants, researchers, seed companies, and chemical industry, has overseen the development, implementation, and evaluation of resistance-management strategies for Bt cotton in Australia (Fitt, 2003). Cotton that produces the Bt toxin Cry1Ac for the control of cotton bollworm (*Helicoverpa armigera*) was replaced by cotton producing Cry1Ac and Cry2Ab in 2004. That allowed producers to reduce the area of refuges from 70 percent with Cry1Ac cotton to as low as 5 percent with Cry1Ac and Cry2Ab cotton (Baker et al., 2008). The exclusive use of a pyramid strategy for managing the evolution of Bt resistance in insect pests might allow producers to use more Bt crops while maintaining efficient resistance management (Mahon et al., 2007; Baker et al., 2008). However, an allele conferring high levels of resistance to Cry2Ab has been found in relatively high frequency (0.0033) in field populations of cotton bollworm (*H. armigera*), and individuals homozygous for this allele can survive on mature cotton producing the toxins Cry1Ac and Cry2Ab (Mahon et al., 2007; Mahon and Olsen, 2009). This indicates that a key assumption of the pyramid strategy is not met (i.e., redundant killing), and thus that caution should be used to manage the evolution of cotton bollworm resistance to Cry1Ac and Cry2Ab cotton in Australia.

Agricultural and Environmental Impacts of Insect Resistance to Bt Crops

The refuge strategy was mandated in the United States not only to slow the evolution of resistance to Bt cultivars but also to protect the effectiveness of Bt sprays. Susceptibility to sprays with many Bt toxins in pests that evolve resistance to single-toxin Bt crops will depend on several factors, including the level of resistance, the variety and concentration of the toxins in the sprays, and the extent of cross-resistance to different toxins. If a spray contains one or more toxins to which the pest has evolved resistance or cross-resistance, susceptibility to the spray could be decreased (Tabashnik et al., 1993; Moar et al., 1995). Nonetheless, sprays containing one or more toxins that kill pests that are resistant to other toxins can be useful against such pests (Tabashnik et al., 1993; Liu et al., 1996; Akhurst et al., 2003; Wang et al., 2007).

Along with other insecticidal compounds with different modes of action, many sprayed Bt insecticides commonly used in the United States contain at least two Cry toxins that differ substantially from each other in amino acid sequence and that bind to different target sites in the larval midgut (Schnepf et al., 1998; Ferré and Van Rie, 2002; Crickmore et al., 2009). The few pests that have evolved resistance to Bt crops could remain susceptible to sprayed Bt insecticides that contain many Bt toxins, as long as the evolution of resistance to the toxins in Bt crops does not involve strong cross-resistance to all the toxins in Bt sprays. Cross-resistance between Bt toxins that differ substantially in amino acid sequence is usually weak or nil, but exceptions occur in important pests that are targeted by Bt crops, including cotton bollworm (*H. zea*) (Hernández-Martínez et al., 2009; Tabashnik et al., 2009b). However, the possibility of cross-resistance in corn stem borer and fall armyworm has not been investigated extensively. The agricultural and environmental impacts of cross-resistance between single-toxin Bt crops and multitoxin Bt sprays will also depend on the extent to which the two approaches are used to control a given pest and on pest movement between Bt crops and areas where Bt sprays are used.

EPA requires remedial action plans to address cases of resistance, which can involve cessation of use of a particular Bt cultivar in a specific area (US-EPA, 1998; Carrière et al., 2001; US-EPA, 2001). Sales of corn that produce the Bt toxin Cry1F were suspended voluntarily in Puerto Rico after the evolution of resistance to Cry1F in fall armyworm (Matten et al., 2008). In the absence of published information on the distribution of resistance and on the presence of cross-resistance and fitness costs, it is not possible to assess whether the evolution of resistance to Cry1F in fall armyworm threatens the sustainability of other Bt crops or sprayed Bt insecticides in Puerto Rico. Furthermore, the economic and environmental consequences of Bt resistance are difficult to assess because little information is available on the profitability of Cry1F corn in Puerto Rico and on how withdrawal of Cry1F corn in Puerto Rico has affected insecticide use.

In contrast, the evolution of resistance of cotton bollworm (*H. zea*) to Cry1Ac cotton in the United States did not have serious agronomic, economic, or environmental consequences. That is because resistance did not affect many bollworm populations, Cry1Ac cotton still provides some control of Cry1Ac-resistant insects, synthetic insecticides were used in conjunction with Cry1Ac cotton from the onset to control bollworm, and the widespread use of cotton that produces both Cry1Ac and Cry2Ab in the states where resistance occurred provided effective control of insects that were resistant to Cry1Ac (Tabashnik et al., 2008). Data on increased bollworm survival on cotton plants that produce Cry1Ac and Cry2Ab in the field or on the consequences of field-evolved resistance to Cry2Ab are lacking (Tabashnik et al., 2009b). Although there is strong evidence of resistance to Cry1Ac in some populations of bollworm in the Southeast (Tabashnik and Carrière, 2009; Tabashnik et al., 2009a), a subset of the data has been contested by some scientists (Moar et al., 2008), and EPA has not commented on the situation.

GENE FLOW AND GENETICALLY ENGINEERED CROPS

This section presents an overview of the potential of gene flow to weedy relatives for crops for which GE varieties have been developed (though not all of these varieties have been commercialized). The movement of herbicide resistance into weedy relatives present on farm fields can influence farmers' weed-management strategies. Gene flow between GE and non-GE crops could accelerate the evolution of pest resistance to Bt crops, if many Bt plants are routinely present in refuges of non-Bt crops (Heuberger et al., 2009; Krupke et al., 2009). The following section specifically considers factors that affect gene flow via cross-pollination within crops, on which the coexistence of GE and non-GE crops depends. Chapter 3 addresses other sources of gene flow, such as co-mingling of seed and germination of volunteer seeds left behind, and the economic consequences of gene flow between GE and non-GE varieties.

Gene Flow Between Genetically Engineered and Non-Genetically Engineered Crops

The potential for cross-pollination between GE and non-GE crops depends on the plant species (Ellstrand, 2003a). In particular, the reproductive strategy of a crop determines the degree of gene flow between GE and non-GE crops; open-pollinated crops (such as corn) have the greatest probability of cross-pollination between GE and non-GE cultivars. Even in self-pollinated plants, out-crossing occurs occasionally, the rate depending on the particular species and environment. In soybean, for example, out-crossing is occasional (Palmer et al., 2001; Abud et al., 2004, 2007). In contrast, corn is freely out-crossing, so the cross-pollination of non-GE cultivars with pollen from GE varieties depends on the distance from the source and other factors. Models of pollen dispersal in corn and the consequent gene flow may have low precision, particularly in light of the small amounts of pollen that move more than 800 ft and the substantial impacts of climate (Ashton et al., 2000).

Among the factors that control gene flow between populations of wind-pollinated plants are distance from the pollen source and pollen shed density, time required for pollen movement, wind speed and direction, air temperature, and relative humidity (Luna V. et al., 2001; Westgate et al., 2003). Pollen viability declines quickly with desiccation. Even if pollen is dispersed over great distances, it may, if dried out, not be viable. Similarly, the occurrence of GE pollen in a non-GE corn field does not necessarily mean that pollination will occur (Feil and Schmid, 2002). Nonetheless, the distances needed to prevent any cross-pollination in corn or other open-pollinated crops are so great that they are not practical in current commercial agricultural systems (Luna V. et al., 2001; Matus-Cádiz et al., 2004).

Insect-mediated cross-pollination between GE and non-GE crops occurs in those species for which insects typically are the agents of transfer of pollen between individuals (Van Deynze et al., 2005; Llewellyn et al., 2007). Canola and cotton are modally out-crossing as a result of pollinator activity, whereas soybean is usually self-pollinated but is visited by insects seeking its pollen (Ahrent and Caviness, 1994; Walklate et al., 2004; Heuberger and Carrière, 2009). In a recent study that monitored cross-pollination of seed-production fields of non-Bt cotton (some HR, some not) by Bt cotton, both the density of flower-foraging honey bees in seed production fields and the area of Bt cotton at a distance of 2,460 ft from the non-Bt cotton fields affected cross-pollination (Heuberger and Carrière, 2009). It had been documented that most cross-pollination in cotton occurs over distances of less than 160 ft (McGregor, 1976; Free, 1993; Xanthopoulos and Kechagia, 2000; Zhang et al., 2005). Nevertheless, foraging honey bees can easily travel two miles or more (Beekman and Ratnieks, 2000); this suggests that the 2,460-ft radius at which pollen from Bt cotton influenced out-crossing of non-Bt cotton resulted from movement of foraging honey bees from Bt to non-Bt cotton fields. Accordingly, the results of Heuberger and Carrière (2009) indicate that small-scale gene-flow studies may miss occasional long-distance cross-pollination between GE and non-GE insect-pollinated crops.

As the adventitious presence of GE traits is widespread in the seed supply of non-GE crops, gene flow between non-GE and GE crops may commonly involve cross-pollination by plants from the same field (i.e., adventitious plants). Heuberger and Carrière (2009) found adventitious Bt cotton plants in 67 percent of seed-production fields of non-Bt cotton. They demonstrated that adventitious Bt plants resulted both from human error (inadvertent planting of Bt cotton in non-Bt cotton fields) and from contamination in seed bags. After accounting for the effect of the area of Bt cotton surrounding a seed-production field and the abundance of foraging honey bees, the density of adventitious Bt plants was positively associated with out-crossing rates in seed-production fields. Most models of pollen transfer between crop varieties have not considered the adventitious presence of GE plants in non-GE fields and may therefore fall short of making accurate predictions of the abundance of GE traits in supposedly non-GE plants.

Canola is not a major crop in the United States, but substantial acreage is planted in North Dakota, which accounted for more than 87 percent of U.S. canola planted in 2009 (USDA-NASS, 2009b). Gene flow between GE and non-GE canola is well documented in the large canola-growing region of western Canada; it can be facilitated by the transient populations of canola established each year outside agricultural fields (Knispel et al., 2008). Genetically engineered HR canola cultivars are the dominant type, and in 2006, glyphosate-resistant and glufosinate-resistant canola cultivars accounted 65 percent and 32 percent, respectively, of U.S. canola acres planted (Howatt, personal communication). Given that there are two GE herbicide-resistance traits (glyphosate and glufosinate) and a non-GE imidazolinone-resistant trait, introgression of these traits can result in multiple-herbicide resistance in a single plant (Knispel et al., 2008). The occurrence of multiple-herbicide–resistant volunteer canola increases the difficulties of management (Beckie et al., 2004; Beckie, 2006; Beckie et al., 2006). In addition to the problem of deploying special management techniques for HR weeds, adventitious presence of a GE trait in a non-GE field of canola has economic consequences.

Alfalfa is an important crop in the United States and is widely cultivated over a broad geographic range (USDA-NASS, 2008). GE glyphosate-resistant alfalfa was commercialized in 2005, and about 198,000 acres were planted in 2006 (Weise, 2007). However, in 2007, it once again became a regulated item, a decision that was upheld by the Court of Appeals in 2008. USDA Animal and Plant Health Inspection Service (APHIS) was ordered to conduct an environmental impact statement (EIS) because of "the significant threat of gene flow and the development of Roundup-resistant weeds that requires further study and analysis in an EIS" (*Geertson Farms v. Johanns*, 2009).[12] APHIS released the draft EIS for public comment in December 2009.

Sugar beet (*Beta vulgaris*) cultivars with the GE trait that confer resistance to glyphosate have been commercialized and were widely adopted by growers in the United States. However, in September 2009, the Northern California District Court ruled that USDA violated the National Environmental Protection Act when it deregulated HR sugar beets, and USDA is required by the court to prepare an environmental impact statement to adequately consider the impacts of GE sugar beets on other sugar beet growers as well as farmers growing table beets and swiss chard, two crops with which sugar beets may cross pollinate (*Center for Food Safety v. Vilsack*, 2009).

---

[12]Roundup is the trademarked name of glyphosate sold by Monsanto.

Gene Flow Between Genetically Engineered
Crops and Related Weed Species

In locations where crop varieties occur with wild or weedy populations of the same or closely related species, interbreeding between a crop and its relatives may lead to exchange of genes between the cultivars or species involved. Such hybridization is common in plants generally and is a key process for the evolution of new plant species. When gene flow occurs between crops and their wild relatives, an agronomic characteristic may move into the wild populations. Environmental sustainability on farms might be affected by the consequences of such gene flow between crops and wild relatives if the gene flow reduces genetic diversity available for crop improvement. However, only a few crops (sunflower, pecan, blueberry, and some squashes) were domesticated within the borders of the United States, so most crops planted on U.S. farms do not pose a risk to the conservation of genetic diversity in related native species and landraces. When crop species coexist with weedy relatives, gene flow might result in a weed-management issue and any accompanying economic and environmental effects. At least 15 crop species have been documented to hybridize with weedy relatives in the United States (Keeler et al., 1996). For HR traits, hybridization is a mechanism by which herbicide resistance might evolve in related weeds if HR crops are able to interbreed with related weedy species occurring in the same location (see the canola example in "Developing Weed-Management Strategies in Herbicide-Resistant Crops" earlier in this chapter).

In the United States for corn and soybean, the most common GE crops grown, no genetically compatible relatives or weedy strains exist; therefore, movement of GE traits into related weed species is not an issue. Wild populations of cotton (*Gossypium hirsutum*) exist along the Gulf Coast and a wild relative (*Gossypium tomentosum*) is in Hawaii. Gene flow is unlikely because the United States prohibits the commercial sale of Bt cotton in those areas. In Hawaii, test varieties and nursery stock can be produced, but with restrictions to minimize gene flow (e.g., US-EPA, 2005). In contrast, the use of HR crops in the same areas is apparently not more restricted than in the rest of the United States (USDA-APHIS, 2008).

Hybridization between the allotetraploid canola (*Brassica napus*) and one of its diploid weedy parents, turnip mustard (*Brassica rapa*) are extensive, and the hybrids are usually about 60 percent pollen fertile (Warwick et al., 2003; Légère, 2005; Simard et al., 2006), thus facilitating the spread of a GE trait into the weeds. GE herbicide-resistant traits have been reported to persist in populations of turnip mustard as they have in several other species of weeds (Warwick et al., 2003; Owen and Zelaya, 2005; York et al., 2005; Warwick et al., 2008). In addition, hybridization is possible between canola and species of a few related genera of mustards, some of them weedy (Warwick et al., 2000; FitzJohn et al., 2007) and occurs spontaneously when they are grown together in an experimental garden. For these reasons, canola has been designated as a moderate-risk crop with regard to the potential for gene flow to its weedy relatives (Stewart et al., 2003), and farm-level effects may occur in canola-growing regions as discussed earlier. The extent to which they have economic impacts and affect environmental sustainability will depend on how the weeds are managed.

Gene flow has been demonstrated from sugar beet to near-relative weeds, *B. macrocarpa* and *B. vulgaris* subsp. *maritima* (Andersen et al., 2005). Thus, the introgression of HR traits from GE sugar beets to weedy beets (*B. vulgaris*) and sea beets (*B. vulgaris* subsp. *maritima*) should be considered a moderate risk (Stewart et al., 2003), but the consequences of gene flow would occur on a very small spatial scale in the United States. Co-occurrence of those species with sugar beet cultivation was limited to two California counties (Kern and Imperial) in 2006 (Calflora, 2009). Weedy beets are sporadic and local in the United States and not considered a major problem, as they are in Europe, where the species is native.

Transgenic virus-resistant squash has been available commercially since 1995 and was estimated to have been planted on an average of 12 percent of the 58,400 acres in squash production in 2005 (Quemada et al., 2008). Wild populations of *Cucurbita pepo* occur in south and central regions of the United States (Cowan and Smith, 1993) and can be an agricultural weed (Oliver et al., 1983). Gene flow between conventional cultivars of domesticated *C. pepo* and its wild populations is known to occur (Kirkpatrick and Wilson, 1988; Decker-Walters et al., 2009). Data are not available yet on the extent to which transgenes may exist in wild populations. As of 2005, the opportunity for gene flow across a large spatial scale appeared low because the majority of transgenic squash cultivation occurs outside the range of the wild populations of *C. pepo* (Quemada et al., 2008). The consequences of any gene flow to wild populations depend on virus incidence and the expression of the transgene in the wild populations (Spencer and Snow, 2001; Fuchs et al., 2004a, 2004b; Laughlin et al., 2009; Sasu et al., 2009)

Concerns about the consequences of a transfer of GE traits to wild or weedy populations and how to effectively mitigate those consequences have delayed the release of GE sunflower (*Helianthus annuus*, a species that was domesticated in the United States), creeping bentgrass (*Agrostis stolonifera*, a popular turf grass introduced from Europe), and rice (*Oryza sativa*, a species native to Asia with related, intercompatible weeds introduced into some rice fields with the crop). For sunflower, the potential for transgene movement to weedy relatives is quite high and thus the consequences of gene flow on weed management presents an environ-

mental concern about the commercial development of transgenic sunflower (Snow and Palma, 1997; Snow, 2002; Ellstrand, 2003b; Stewart et
al., 2003). Wild sunflowers are weeds in row-crop fields, including corn,
soybean, domesticated sunflower, wheat, and small grains (Bernards et
al., 2009). In a multiyear study conducted across the High Plains of the
United States in a number of commercial sunflower-production fields,
it was observed that approximately 66 percent of the fields existed near
weedy sunflower (Burke et al., 2002). It is important that the cultivated
and weedy sunflowers flowered simultaneously (52–96 percent), and evidence of hybridization ranged from 10 percent to 33 percent of the weedy
sunflower (Burke et al., 2002). Evidence of genes coding for herbicide
resistance in cultivated sunflower moving to weedy sunflower suggests
that there is a substantial risk of the introgression of the trait into wild
sunflower populations, which might result in increased management
problems for growers if the wild plants are in cultivated fields (Marshall,
2001; Massinga and Al-Khatib, 2002; Massinga, 2003). These management problems may emerge in association with sunflower varieties with
resistance to the herbicide imazamox that were developed using conventional breeding methods and not genetic engineering. Hybrids of
imazamox-resistant sunflowers and two interbreeding relatives appear
to be competitively equal to the HR domesticated sunflower, suggesting
that the resistance gene will persist when gene flow occurs (Massinga et
al., 2005).

GE glyphosate-resistant creeping bentgrass was field-tested in Oregon
in 2003, and introgression of the transgene into weedy populations was
detected at a considerable distance from the test sites (Mallory-Smith et
al., 2005; Reichman et al., 2006). The inability to mitigate GE trait introgression into compatible weedy relatives of creeping bentgrass has delayed
commercialization of the GE creeping bentgrass product (Charles, 2007).
The transfer of this trait may be important if glyphosate is used to control
weedy populations of the grass.

In the case of rice, GE glufosinate-resistant rice cultivars have been
developed to improve weed management of red rice (*Oryza sativa* L.), a
common and important weed in commercial rice production. However,
these GE cultivars have never been commercially available (Gealy et al.,
2007) though they were first deregulated in 1998. The commercialization of rice with resistance to the herbicide imazethapyr (produced by a
chemical-induced seed mutagenesis, not genetic engineering) has resulted
in the movement of this HR trait into red rice because total control of
red rice is not always possible (Burgos et al., 2008). Stewardship recommendations to prevent and to control the spread of HR red rice exist and
encourage crop diversity (BASF, 2009); data on the success of managing
imazethapyr-resistant red rice may provide useful information for the

regulation of GE cultivars. In addition to concerns about introgression with the weed red rice, GE rice has historically been unacceptable to consumers for the reasons discussed in Chapter 1 (Gealy and Dilday, 1997; Gealy et al., 2003, 2007).

Wheat is a major grain crop in the United States, and there has been interest in commercializing genetically engineered HR cultivars. The potential for introgression of a GE trait into near-relative weed populations exists (Ellstrand et al., 1999; Morrison et al., 2002a, 2002b; Stewart et al., 2003). Jointed goatgrass (*Aegilops cylindrica*) is reported to be an important weed of small grains in Colorado, Kansas, New Mexico, Oklahoma, Oregon, Utah, Washington, and Wyoming (NAPPO, 2003) and is a weed in the Great Plains (Stubbendieck et al., 1994). Specifically, it causes serious problems in winter wheat in the western United States due to its similarity to wheat in appearance, seed size, growth pattern, and genetics (Schmale et al., 2008, 2009; Yenish et al., 2009). Studies have demonstrated hybridization with wheat varieties (Hanson et al., 2005; Rehman et al., 2006). It is predicted that hybridization between HR wheat cultivars and *Aegilops* spp. will result in the introgression of the HR trait and thus a more competitive weedy hybrid, complicating management issues in wheat production (Hanson et al., 2005; Loureiro et al., 2008). Glyphosate-resistant hard red spring wheat provided an opportunity for better weed management and resulted in a 10-percent higher grain yield than conventional wheat cultivars treated with conventional herbicides (Howatt et al., 2006). Despite the technical fit, the program to develop glyphosate-resistant cultivars was postponed in 2004 because of regulatory and marketing issues, concern for stewardship, and the inability to ensure the segregation of the GE wheat from non-GE wheat grain at the time (Dill, 2005) (see Chapter 4 for further details). An imazamox-resistant winter wheat, which was bred using conventional techniques, has been commercially available since 2002, and the identical concerns about the development of HR jointed goatgrass biotypes exist (Kniss et al., 2008).

The ecological and economic consequences of the introgression of GE traits into weedy or native relatives will vary among types of GE traits (Owen, 2008). For HR crops, the introgression of herbicide resistance from GE crops into weedy near-relatives is likely to have consequences for weed management *when* weeds with the resistance trait occur in fields or other ecosystems treated with the herbicide. Therefore, for future HR plants, understanding the extent to which a herbicide controls weedy relatives will provide valuable information on the consequences of gene flow for on-farm and off-farm weed management.

CONCLUSIONS

Environmental effects at the farm level have occurred as a result of the adoption of GE crops and the agricultural practices that accompany their cultivation. The introduction of GE crops has reduced pesticide use or the toxicity of pesticides used on fields where soybean, corn, and cotton are grown. Available evidence indicates that no-till practices and HR crops are complementary, and each has encouraged the other's adoption. Conservation tillage, especially no-till, reduces soil erosion and can improve soil quality. The pesticide shifts and increase in conservation tillage with GE crops have generally benefited farmers who adopted them so far. Conservation tillage practices can also improve water quality by reducing the volume of runoff from farms into surface water, thereby reducing sedimentation and contamination from farm chemicals. Given that agriculture is the largest cause of impaired quality of surface waters, that may constitute the largest benefit of GE crops, but the infrastructure for tracking and understanding this does not exist.

The effects of Bt crops on nontarget invertebrates, including predators, are favorable or neutral, depending on the degree to which Bt crops replace insecticide treatment and on whether additional insecticide treatments are applied to the Bt crop. Evidence indicates no effect of Bt toxins on the honey bee, a widespread pollinator in agricultural systems. For HR crops, the effects on the abundance of arthropods in the fields correlate with whether weeds are controlled more effectively. Shifts in the weed communities have occurred in response to weed-management tactics used for HR crops, in particular when weeds in glyphosate-resistant crops are treated only with glyphosate. Similarly, glyphosate-resistant weeds have evolved where the glyphosate application is repeated and constitutes the only weed-management tactic used. The evolution of resistance to glyphosate in particular kinds of weeds and shifts in the weed community may increase production costs for farmers, require more tillage for weed control, and lead to at least a partial return to the use of different and often more toxic herbicides. The development and establishment of more diversified control strategies for managing weeds in HR crops is needed.

The first generation of IR crops commercialized in the United States produced a single Bt toxin for the control of insect pests. Since the commercialization of those crops, EPA has mandated the refuge strategy to delay the evolution of resistance in major insect pests that are controlled by Bt corn and cotton. After 14 years of use of Bt crops, two insect pests have evolved resistance to Bt crops in the United States: Cotton bollworm (*Helicoverpa zea*) evolved resistance to Cry1Ac and Cry2Ab in Bt cotton, and fall armyworm evolved resistance to Cry1F in Bt corn. The evolution in bollworm of resistance to Bt cotton did not have serious agronomic, economic, or environmental consequences. Information for assessing the consequence of the evolution of resistance to Cry1F in Bt corn in fall army-worm is lacking. The second generation of IR crops produces two or more Bt toxins for the control of individual insect pest species. The complete replacement of one-toxin with multitoxin Bt crops should help in delaying the evolution of insect-pest resistance to IR crops.

The changes in weed and insect population densities resulting from the adoption of HR and IR crops can affect farms beyond the boundaries of the operations that are using the GE crops. That is, farming practices may have landscape as well as local level effects on pest populations. For example, large-scale planting of IR crops has decreased populations of some insect pests targeted by Bt crops not just at a farm-field level but on a regional scale. It can also affect local and possibly landscape populations of nontarget or beneficial organisms according to crop species planted and management of pests, nutrients, water, and soil (Björklund et al., 1999; Cattaneo et al., 2006; Dale and Polasky, 2007; Zhang et al., 2007; Carrière et al., 2009). Beneficial organisms with high mobility move among habitats and crop fields, so effects within a field that is planted to GE crops could influence beneficial organisms on other farms as well as noncultivated habitats in the region.

For corn and soybean, gene flow between GE varieties and wild relatives is not an issue in the United States because corn and soybean have no wild relatives here. Limited overlap occurs between cotton and wild relatives and between sugar beet and introduced, weedy relatives. Gene flow is unlikely for Bt cotton and wild relatives because of planting restrictions, but there are no planting restrictions for HR cotton. Other crops in which gene flow with wild or weedy relatives is possible include canola, alfalfa, sunflower, creeping bentgrass, wheat, and rice. Gene flow between GE and non-GE crops occurs via cross-pollination between GE and non-GE plants from different fields, co-mingling of seed before or during the production year, and germination of seeds that are left behind after the production year. Gene flow between GE and non-GE crops is almost impossible to prevent completely with current technology. The complex interactions among the multiple factors that influence gene flow between GE and non-GE crops and the resulting levels of adventitious presence of GE traits in non-GE crops deserve more attention.

REFERENCES

Abud, S., P.I.M. Souza, C.T. Moreira, A.L. Farias Neto, G.R. Vianna, S.R.M. Andrade, J. Nunes Junior, R.A. Guerzoni, P.M.F.O. Monteiro, M.S. Assuncao, E.L. Rech, and F.J.L. Aragao. 2004. Distance of flow in transgenic BROO-69515 and the non transgenic soybean in the Cerrado Region, Brasil. In *Proceedings of VII World Soybean Research Conference*, pp. 110–111. eds. F. Moscardi, C.B. Hoffmann-Campo, O.F. Saraiva, P.R. Galerani, F.C. Krzyzanowski, and M.C. Carrao-Panizzi. Foz do Iguassu, Brazil: Brasilian Agricultural Research Corporation.

Abud, S., P.I.M. De Souza, G.R. Vianna, E. Leonardecz, C.T. Moreira, F.G. Faleiro, J.N. Júnior, P.M.F.O. Monteiro, E.L. Rech, and F.J.L. Aragão. 2007. Gene flow from transgenic to nontransgenic soybean plants in the Cerrado region of Brazil. *Genetics and Molecular Research* 6(2):445–452.

Adamczyk, J.J., Jr., L.C. Adams, and D.D. Hardee. 2001. Field efficacy and seasonal expression profiles for terminal leaves of single and double *Bacillus thuringiensis* toxin cotton genotypes. *Journal of Economic Entomology* 94(6):1589–1593.

Ahl Goy, P., G. Warren, J. White, L. Privalle, P. Fearing, and D. Vlachos. 1995. Interaction of an insect tolerant maize with organisms in the ecosystem. *Mitteilungen aus der Biologischen Bundesanstalt fuer Land-und Forstwirtschaft Berlin-Dahlem* (309):50–53.

Ahrent, D.K., and C.E. Caviness. 1994. Natural cross-pollination of twelve soybean cultivars in Arkansas. *Crop Science* 34(2):376–378.

Akhurst, R.J., W. James, L.J. Bird, and C. Beard. 2003. Resistance to the Cry1Ac δ-endotoxin of *Bacillus thuringiensis* in the cotton bollworm, *Helicoverpa armigera* (Lepidoptera: Noctuidae). *Journal of Economic Entomology* 96:1290–1299.

Al-Deeb, M.A., G.E. Wilde, J.M. Blair, and T.C. Todd. 2003. Effect of Bt corn for corn rootworm control on nontarget soil microarthropods and nematodes. *Environmental Entomology* 32(4):859–865.

Alstad, D.N., and D.A. Andow. 1995. Managing the evolution of insect resistance to transgenic plants. *Science* 268(5219):1894–1896.

Andersen, N.S., H.R. Siegismund, V. Meyer, and R.B. Jørgensen. 2005. Low level of gene flow from cultivated beets (*Beta vulgaris* L. ssp. *vulgaris*) into Danish populations of sea beet (*Beta vulgaris* L. ssp. *maritima* (L.) Arcangeli). *Molecular Ecology* 14(5):1391–1405.

Andow, D.A., D.M. Olson, R.L. Hellmich, D.N. Alstad, and W.D. Hutchison. 2000. Frequency of resistance to *Bacillus thuringiensis* toxin Cry1Ab in an Iowa population of European corn borer (Lepidoptera: Crambidae). *Journal of Economic Entomology* 93(1):26–30.

Andow, D.A., and A.R. Ives. 2002. Monitoring and adaptive resistance management. *Ecological Applications* 12(5):1378–1390.

APRD (Arthropod Pesticide Resistance Database). 2009. eds. E. Whalon, D. Mota-Sanchez, R.M. Hollingworth, and L. Duynslager. Available online at http://www.pesticideresistance.com/. Accessed December 21, 2009.

Ashton, B.A., D.S. Ireland, and M.E. Westgate. 2000. Applicability of the industrial source complex (ISC) air dispersion model for use on corn pollen transport. B.Sc. thesis. Meteorology Program, Iowa State University, Ames, IA.

Bagwell, R.D., D.R. Cook, B.R. Leonard, S. Micinski, and E. Burris. 2001. Status of insecticide resistance in tobacco bollworm and bollworm in Louisiana during 2000. In *Proceedings of the Beltwide Cotton Conference*, pp. 785–789 (Memphis, TN, January 10–11, 2001). National Cotton Council of America.

Baker, G.H., C.R. Tann, and G.P. Fitt. 2008. Production of *Helicoverpa* spp. (Lepidoptera, Noctuidae) from different refuge crops to accompany transgenic cotton plantings in eastern Australia. *Australian Journal of Agricultural Research* 59(8):723–732.

Baker, H.G. 1991. The continuing evolution of weeds. *Economic Botany* 45:445–449.

BASF (Baden Aniline and Soda Factory). 2009. *CLEARFIELD® rice stewardship recommendations*. APN 08-10-830-0009. BASF Corporation. Research Triangle Park, NC. Available online at http://agproducts.basf.us/products/research-library/clearfield-rice-stewardship.pdf. Accessed November 11, 2009.

Baucom, R.S., and R. Mauricio. 2004. Fitness costs and benefits of novel herbicide tolerance in a noxious weed. *Proceedings of the National Academy of Sciences of the United States of America* 101(36):13386–13390.

Baumgarte, S., and C.C. Tebbe. 2005. Field studies on the environmental fate of the Cry1Ab Bt-toxin produced by transgenic maize (MON810) and its effect on bacterial communities in the maize rhizosphere. *Molecular Ecology* 14(8):2539–2551.

Beckie, H.J. 2006. Herbicide-resistant weeds: Management tactics and practices. *Weed Technology* 20(3):793–814.

Beckie, H.J., S.I. Warwick, H. Nair, and G. Séguin-Swartz. 2003. Gene flow in commercial fields of herbicide-resistant canola (*Brassica napus*). *Ecological Applications* 13(5):1276–1294.

Beckie, H.J., G. Séguin-Swartz, H. Nair, S.I. Warwick, and E. Johnson. 2004. Multiple herbicide-resistant canola can be controlled by alternative herbicides. *Weed Science* 52(1):152–157.

Beckie, H.J., K.N. Harker, L.M. Hall, S.I. Warwick, A. Légère, P.H. Sikkema, G.W. Clayton, A.G. Thomas, J.Y. Leeson, G. Séguin-Swartz, and M.J. Simard. 2006. A decade of herbicide-resistant crops in Canada. *Canadian Journal of Plant Science* 86(4):1243–1264.

Beekman, M., and F.L.W. Ratnieks. 2000. Long-range foraging by the honey-bee, *Apis mellifera* L. *Functional Ecology* 14(4):490–496.

Behrens, M.R., N. Mutlu, S. Chakraborty, R. Dumitru, Z.J. Wen, B.J. LaVallee, P.L. Herman, T.E. Clemente, and D.P. Weeks. 2007. Dicamba resistance: Enlarging and preserving biotechnology-based weed management strategies. *Science* 316(5828):1185–1188.

Bernards, M.L., R.E. Gaussoin, R.N. Klein, S.Z. Knezevic, D.J. Lyon, R.G. Wilson, P.J. Shea, and C.L. Ogg. 2009. 2009 Guide for weed management in Nebraska. EC130. University of Nebraska-Lincoln. Lincoln, NE. Available online at http://www.ianrpubs.unl.edu/sendIt/ec130.pdf. Accessed September 18, 2009.

Berwald, D., S. Matten, and D. Widawsky. 2006. Economic analysis and regulating pesticide biotechnology at the U.S. Environmental Protection Agency. In *Regulating agricultural biotechnology: Economics and policy*. eds. R.E. Just, J.M. Alston, and D. Zilberman, pp. 21–36. New York: Springer.

Birch, A.N.E., B.S. Griffiths, S. Caul, J. Thompson, L.H. Heckmann, P.H. Krogh, and J. Cortet. 2007. The role of laboratory, glasshouse and field scale experiments in understanding the interactions between genetically modified crops and soil ecosystems: A review of the ECOGEN project. *Pedobiologia* 51(3):251–260.

Björklund, J., K.E. Limburg, and T. Rydberg. 1999. Impact of production intensity on the ability of the agricultural landscape to generate ecosystem services: An example from Sweden. *Ecological Economics* 29(2):269–291.

Blanco-Canqui, H., and R. Lal. 2009. Crop residue removal impacts on soil productivity and environmental quality. *Critical Reviews in Plant Sciences* 28(3):139–163.

Boerboom, C.M. 1999. Nonchemical options for delaying weed resistance to herbicides in Midwest cropping system. *Weed Technology* 13:636–642.

Boerboom, C. 2005. Glyphosate-resistant common lambsquarters announced. *Wisconsin Crop Manager* 12(1):8–9.

Bommireddy, P.L., and B.R. Leonard. 2008. Age-specific mortality of *Helicoverpa zea* and *Heliothis virescens* (Lepidoptera: Noctuidae) larvae on flower buds of transgenic cotton expressing Vip3a and Vipcot$^{(TM)}$ insecticidal proteins. *Journal of Entomological Science* 43(4):349–361.

Borggaard, O.K., and A.L. Gimsing. 2008. Fate of glyphosate in soil and the possibility of leaching to ground and surface waters: A review. *Pest Management Science* 64(4):441–456.

Bourguet, D., J. Chaufaux, A. Micoud, M. Delos, B. Naibo, F. Bombarde, G. Marque, N. Eychenne, and C. Pagliari. 2002. *Ostrinia nubilalis* parasitism and the field abundance of non-target insects in transgenic *Bacillus thuringiensis* corn (*Zea mays*). *Environmental Biosafety Research* 1(1):49–60.

Bravo, A., and M. Soberón. 2008. How to cope with insect resistance to Bt toxins? *Trends in Biotechnology* 26(10):573–579.

Buhler, D.D. 1992. Population dynamics and control of annual weeds in corn (*Zea mays*) as influenced by tillage systems. *Weed Science* 40:241–248.

Buhler, D.D., and M.D.K. Owen. 1997. Emergence and survival of horseweed (*Conyza canadensis*). *Weed Science* 45:98–101.

Buhler, D.D., R.G. Hartzler, and F. Forcella. 1997. Implications of weed seedbank dynamics to weed management. *Weed Science* 45:329–336.

Burgos, N.R., J.K. Norsworthy, R.C. Scott, and K.L. Smith. 2008. Red rice (*Oryza sativa*) status after 5 years of imidazolinone-resistant rice technology in Arkansas. *Weed Technology* 22(1):200–208.

Burke, J.M., K.A. Gardner, and L.H. Rieseberg. 2002. The potential for gene flow between cultivated and wild sunflower (*Helianthus annuus*) in the United States. *American Journal of Botany* 89(9):1550–1552.

Calflora. 2009. Information on California plants for education, research and conservation. The Calflora Database. Available online at http://www.calflora.org/cgi-bin/specieslist. cgi?orderby=taxon&where-genus=Beta&ttime=1251561416&ttime=1251561417. Accessed August 29, 2009.

Carlisle, S.M., and J.T. Trevors. 1986. Effect of the herbicide glyphosate on respiration and hydrogen consumption in soil. *Water, Air, and Soil Pollution* 27(3–4):391–401.

Carrière, Y., and D.A. Roff. 1995. Change in genetic architecture resulting from the evolution of insecticide resistance: A theoretical and empirical analysis. *Heredity* 75(6):618–629.

Carrière, Y., and B.E. Tabashnik. 2001. Reversing insect adaptation to transgenic insecticidal plants. *Proceedings of the Royal Society B: Biological Sciences* 268(1475):1475–1480.

Carrière, Y., T.J. Dennehy, B. Pedersen, S. Haller, C. Ellers-Kirk, L. Antilla, Y.B. Liu, E. Willott, and B.E. Tabashnik. 2001. Large-scale management of insect resistance to transgenic cotton in Arizona: Can transgenic insecticidal crops be sustained? *Journal of Economic Entomology* 94(2):315–325.

Carrière, Y., C. Ellers-Kirk, M.S. Sisterson, L. Antilla, M. Whitlow, T.J. Dennehy, and B.E. Tabashnik. 2003. Long-term regional suppression of pink bollworm by *Bacillus thuringiensis* cotton. *Proceedings of the National Academy of Sciences of the United States of America* 100(4):1519–1523.

Carrière, Y., P. Dutilleul, C. Ellers-Kirk, B. Pedersen, S. Haller, L. Antilla, T.J. Dennehy, and B.E. Tabashnik. 2004a. Sources, sinks, and the zone of influence of refuges for managing insect resistance to Bt crops. *Ecological Applications* 14(6):1615–1623.

Carrière, Y., M.S. Sisterson, and B.E. Tabashnik. 2004b. Resistance management for sustainable use of Bacillus thuringiensis crops in integrated pest management. In *Insect pest management: Field and protected crops*. eds. A.R. Horowitz and I. Ishaaya, pp. 65–95. Berlin: Springer.

Carrière, Y., C. Ellers-Kirk, K. Kumar, S. Heuberger, M. Whitlow, L. Antilla, T.J. Dennehy, and B.E. Tabashnik. 2005. Long-term evaluation of compliance with refuge requirements for Bt cotton. *Pest Management Science* 61(4):327–330.

Carrière, Y., C. Ellers-Kirk, M.G. Cattaneo, C.M. Yafuso, L. Antilla, C.-Y. Huang, M. Rahman, B.J. Orr, and S.E. Marsh. 2009. Landscape effects of transgenic cotton on non-target ants and beetles. *Basic and Applied Ecology* 10(7):597–606.

Cattaneo, M.G., C.M. Yafuso, C. Schmidt, C.-Y. Huang, M. Rahman, C. Olson, C. Ellers-Kirk, B.J. Orr, S.E. Marsh, L. Antilla, P. Dutilleul, and Y. Carrière. 2006. Farm-scale evaluation of the impacts of transgenic cotton on biodiversity, pesticide use, and yield. *Proceedings of the National Academy of Sciences of the United States of America* 103(20):7571–7576.

Center for Food Safety, et al. v. Thomas J. Vilsack, et al.: Order regarding cross-motions for summary judgment. 2009. U.S. District Court for the Northern District of California. C 08-00484 JSW, Case #: Case3:08-cv-00484-JSW. Available online at http://www.earthjustice.org/library/legal_docs/9-21-09-order.pdf. Accessed December 16, 2009.

Cerdeira, A.L., and S.O. Duke. 2006. The current status and environmental impacts of glyphosate-resistant crops: A review. *Journal of Environmental Quality* 35(5):1633–1658.

Charles, D. 2007. Environmental regulation. U.S. courts say transgenic crops need tighter scrutiny. *Science* 315(5815):1069.

Clark, B.W., K.R. Prihoda, and J.R. Coats. 2006. Subacute effects of transgenic Cry1Ab *Bacillus thuringiensis* corn litter on the isopods *Trachelipus rathkii* and *Armadillidium nasatum*. *Environmental Toxicology and Chemistry* 25(10):2653–2661.

Cortet, J., M.N. Andersen, S. Caul, B. Griffiths, R. Joffre, B. Lacroix, C. Sausse, J. Thompson, and P.H. Krogh. 2006. Decomposition processes under Bt (*Bacillus thuringiensis*) maize: Results of a multi-site experiment. *Soil Biology and Biochemistry* 38(1):195–199.

Cowan, C.W., and B.D. Smith. 1993. New perspectives on a wild gourd in eastern North America. *Journal of Ethnobiology* 13(1):17–54.

Crickmore, N., D.R. Zeigler, J. Feitelson, E. Schnepf, J.V. Rie, D. Lereclus, J. Baum, and D.H. Dean. 2009. *Bacillus thuringiensis* toxin nomenclature. Brighton, UK: University of Sussex. Available online at http://www.lifesci.sussex.ac.uk/home/Neil_Crickmore/Bt/. Accessed June 26, 2009.

Crowder, D.W., and Y. Carrière. 2009. Comparing the refuge strategy for managing the evolution of insect resistance under different reproductive strategies. *Journal of Theoretical Biology* 261(3):423–430.

CTIC (Conservation Technology Information Center). 2009. National crop residue management survey. Available online at http://www.conservationinformation.org/?action=members_crmsurvey. Accessed October 1, 2009.

Culpepper, A.S. 2006. Glyphosate-induced weed shifts. *Weed Technology* 20(2):277–281.

Culpepper, A.S., and A.C. York. 2007. Glyphosate-resistant Palmer amaranth impacts southeastern agriculture. In *Proceedings of the Illinois Crop Protection Technology Conference*, pp. 61-63 (Champaign, IL, January 3–4, 2007). University of Illinois Urbana-Champaign. Available online at http://ipm.illinois.edu/education/proceedings/icptcp2007.pdf. Accessed June 18, 2009.

Dale, V.H., and S. Polasky. 2007. Measures of the effects of agricultural practices on ecosystem services. *Ecological Economics* 64(2):286–296.

Davies, T.G.E., L.M. Field, P.N.R. Usherwood, and M.S. Williamson. 2007. DDT, pyrethrins, pyrethroids and insect sodium channels. *IUBMB Life* 59(3):151–162.

Decker-Walters, D.S., J.E. Staub, S.-M. Chung, E. Nakata, and H.D. Quemada. 2009. Diversity in free-living populations of *Cucurbita pepo* (Cucurbitaceae) as assessed by random amplified polymorphic DNA. *Systematic Botany* 27(1):19–28.

Devare, M., L.M. Londoño-R, and J.E. Thies. 2007. Neither transgenic Bt maize (MON863) nor tefluthrin insecticide adversely affect soil microbial activity or biomass: A 3-year field analysis. *Soil Biology and Biochemistry* 39(8):2038–2047.

Devare, M.H., C.M. Jones, and J.E. Thies. 2004. Effect of Cry3Bb transgenic corn and tefluthrin on the soil microbial community: Biomass, activity, and diversity. *Journal of Environmental Quality* 33(3):837–843.

Dill, G.M. 2005. Glyphosate-resistant crops: History, status and future. *Pest Management Science* 61(3):219–224.

Donegan, K.K., C.J. Palm, V.J. Fieland, L.A. Porteous, L.M. Ganio, D.L. Schaller, L.Q. Bucao, and R.J. Seidler. 1995. Changes in levels, species and DNA fingerprints of soil microorganisms associated with cotton expressing the *Bacillus thuringiensis* var. *kurstaki* endotoxin. *Applied Soil Ecology* 2(2):111–124.

Duan, J.J., M. Marvier, J. Huesing, G. Dively, and Z.Y. Huang. 2008. A meta-analysis of effects of Bt crops on honey bees (Hymenoptera: Apidae). *PLoS ONE* 3(1).

Duke, S.O. 2005. Taking stock of herbicide-resistant crops ten years after introduction. *Pest Management Science* 61(3):211–218.

Dutton, A., H. Klein, J. Romeis, and F. Bigler. 2002. Uptake of Bt-toxin by herbivores feeding on transgenic maize and consequences for the predator *Chrysoperla carnea*. *Ecological Entomology* 27(4):441–447.

Ellstrand, N.C. 2003a. Current knowledge of gene flow in plants: Implications for transgene flow. *Philosophical Transactions of the Royal Society B: Biological Sciences* 358(1434):1163–1170.

———. 2003b. *Dangerous liaisons? When cultivated plants mate with their wild relatives.* Baltimore, MD: Johns Hopkins University Press.

Ellstrand, N.C., H.C. Prentice, and J.F. Hancock. 1999. Gene flow and introgression from domesticated plants into their wild relatives. *Annual Review of Ecology and Systematics* 30(1):539–563.

Estes, T.L., R. Allen, R.L. Jones, D.R. Buckler, K.H. Carr, D.I. Gustafson, C. Custin, M.J. McKee, A.G. Hornsby, and R.P. Richards. 2001. Predicted impact of transgenic crops on water quality and related ecosystems in vulnerable watersheds of the United States. In *Pesticide behaviour in soils and water: Proceedings of a symposium organised by the British Crop Protection Council*, pp. 357–366 (Brighton, UK, November 13–15, 2001). Farmham, Surrey, UK: British Crop Protection Council.

FAO (Food and Agriculture Organization). 2008. FAOStat: ResourceSTAT. FAO of the United Nations. Available online at http://faostat.fao.org. Accessed June 16, 2009.

Fawcett, R.S., B.R. Christensen, and D.P. Tierney. 1994. The impact of conservation tillage on pesticide runoff into surface water: A review and analysis. *Journal of Soil and Water Conservation* 49(2):126–135.

Feil, B., and J.E. Schmid. 2002. *Dispersal of maize, wheat and rye pollen: A contribution to determining the necessary isolation distances for the cultivation of transgenic crops.* Aachen, Germany: Shaker Verlag GmbH.

Fernandez-Cornejo, J., and W.D. McBride. 2002. Adoption of bioengineered crops. Agricultural Economic Report No. 810. May 1. U.S. Department of Agriculture–Economic Research Service. Washington, DC. Available online at http://www.ers.usda.gov/publications/aer810/. Accessed July 25, 2009.

Fernandez-Cornejo, J., and M.F. Caswell. 2006. The first decade of genetically engineered crops in the United States. Economic Information Bulletin No. 11. April. U.S. Department of Agriculture–Economic Research Service. Washington, DC. Available online at http://www.ers.usda.gov/publications/eib11/eib11.pdf. Accessed June 15, 2009.

Fernandez-Cornejo, J., C. Klotz-Ingram, R. Heimlich, M. Soule, W. McBride, and S. Jans. 2003. Economic and environmental impacts of herbicide tolerant and insect resistant crops in the United States. In *The economic and environmental impacts of agbiotech: A global perspective.* ed. N.G. Kalaitzandonakes, pp. 63–88. New York: Kluwer Academic/Plenum Publishers.

Fernandez-Cornejo, J., R. Nehring, E.N. Sinha, A. Grube, and A. Vialou. 2009. Assessing recent trends in pesticide use in U.S. agriculture. Paper presented at the 2009 Annual Meeting of the Agricultural and Applied Economics Association (Milwaukee, WI, July 26–28, 2009). Available online at http://ageconsearch.umn.edu/handle/49271. Accessed October 1, 2009.

Ferré, J., and J. Van Rie. 2002. Biochemistry and genetics of insect resistance to *Bacillus thuringiensis. Annual Review of Entomology* 47(1):501–533.

Fitt, G.P. 2003. Deployment and impact of transgenic Bt cottons in Australia. In *The economic and environmental impacts of agbiotech : A global perspective.* ed. N.G. Kalaitzandonakes, pp. 141–164. New York: Kluwer Academic/Plenum Publishers.

FitzJohn, R., T. Armstrong, L. Newstrom-Lloyd, A. Wilton, and M. Cochrane. 2007. Hybridisation within *Brassica* and allied genera: Evaluation of potential for transgene escape. *Euphytica* 158(1):209–230.

Flores, S., D. Saxena, and G. Stotzky. 2005. Transgenic Bt plants decompose less in soil than non-Bt plants. *Soil Biology and Biochemistry* 37(6):1073–1082.

Folmar, L.C., H.O. Sanders, and A.M. Julin. 1979. Toxicity of the herbicide glyphosate and several of its formulations to fish and aquatic invertebrates. *Archives of Environmental Contamination and Toxicology* 8(3):269–278.

Foresman, C., and L. Glasgow. 2008. US grower perceptions and experiences with glyphosate-resistant weeds. *Pest Management Science* 64(4):388–391.

Foster, G., and S. Dabney. 1995. Agricultural tillage systems: Water erosion and sedimentation. In *Farming for a better environment: A white paper*. Ankeny, Iowa: Soil and Water Conservation Society.

Free, J.B. 1993. *Insect pollination of crops*. 2nd ed. San Diego, CA: Academic Press.

Frisvold, G., A. Boor, and J.M. Reeves. 2007. Simultaneous diffusion of herbicide tolerant cotton and conservation tillage. In *Proceedings of the 2007 Beltwide Cotton Conference* (New Orleans, LA, January 9–12, 2007). National Cotton Council of America.

Fuchs, M., E.M. Chirco, and D. Gonsalves. 2004a. Movement of coat protein genes from a commercial virus-resistant transgenic squash into a wild relative. *Environmental Biosafety Research* 3(1):5–16.

Fuchs, M., E.M. Chirco, J.R. McFerson, and D. Gonsalves. 2004b. Comparative fitness of a wild squash species and three generations of hybrids between wild × virus-resistant transgenic squash. *Environmental Biosafety Research* 3(1):17–28.

Gahan, L.J., F. Gould, and D.G. Heckel. 2001. Identification of a gene associated with Bt resistance in *Heliothis virescens*. *Science* 293(5531):857–860.

Gassmann, A.J., Y. Carrière, and B.E. Tabashnik. 2009. Fitness costs of insect resistance to *Bacillus thuringiensis*. *Annual Review of Entomology* 54(1):147–163.

Gealy, D.R., and R.H. Dilday. 1997. Biology of red rice (*Oryza sativa* L.) accessions and their susceptibility to glufosinate and other herbicide. Abstract. *Weed Science Society of America* 37:34.

Gealy, D.R., D.H. Mitten, and J.N. Rutger. 2003. Gene flow between red rice (*Oryza sativa*) and herbicide-resistant rice (*O. sativa*): Implications for weed management. *Weed Technology* 17(3):627–645.

Gealy, D.R., K.J. Bradford, L. Hall, R. Hellmich, A. Raybould, J. Wolt, and D. Zilberman. 2007. Implications of gene flow in the scale-up and commercial use of biotechnology-derived crops: Economic and policy considerations. Council for Agricultural Science and Technology, December (Issue 37). Available online at http://www.cast-science.org/websiteUploads/publicationPDFs/CAST%20Issue%20Paper%2037%20galley-final-2149.pdf. Accessed October 21, 2007.

Geertson Farms Inc., et al. v. Mike Johanns, et al., and Monsanto Company: Memorandum and order Re: Permanent injunction. 2009. U.S. District Court for the Northern District of California. C 06-01075 CRB, Case #: 3:06-cv-01075-CRB. Decided May 3, 2007. Available online at http://www.aphis.usda.gov/brs/pdf/Alfalfa_Ruling_20070503.pdf. Accessed December 16, 2009.

Georghiou, G.P. 1986. The magnitude of the resistance problem. In *Pesticide resistance: Strategies and tactics for management*, pp. 14–43. Washington, DC: National Academy Press.

Giesy, J.P., S. Dobson, and K.R. Solomon. 2000. Ecotoxicological risk assessment for Roundup® herbicide. *Reviews of Environmental Contamination and Toxicology* 167:35–120.

Givnish, T.J. 2001. The rise and fall of plant species: A population biologist's perspective. *American Journal of Botany* 88(10):1928–1934.

Glare, T.R., and M. O'Callaghan. 2000. *Bacillus thuringiensis: Biology, ecology, and safety*. New York: Wiley.

Gould, F. 1995. Comparisons between resistance management strategies for insects and weeds. *Weed Technology* 9(4):830–839.

Gould, F. 1998. Sustainability of transgenic insecticidal cultivars: Integrating pest genetics and ecology. *Annual Review of Entomology* 43:701–726.

————. 2003. Bt-resistance management—theory meets data. *Nature Biotechnology* 21(12):1450–1451.

Gould, F., A. Anderson, A. Jones, D. Sumerford, D.G. Heckel, J. Lopez, S. Micinski, R. Leonard, and M. Laster. 1997. Initial frequency of alleles for resistance to *Bacillus thuringiensis* toxins in field populations of *Heliothis virescens. Proceedings of the National Academy of Sciences of the United States of America* 94(8):3519–3523.

Gould, F., M.B. Cohen, J.S. Bentur, G.G. Kennedy, and J. Van Duyn. 2006. Impact of small fitness costs on pest adaptation to crop varieties with multiple toxins: A heuristic model. *Journal of Economic Entomology* 99(6):2091–2099.

Green, J.M. 2007. Review of glyphosate and ALS-inhibiting herbicide crop resistance and resistant weed management. *Weed Technology* 21(2):547–558.

Green, J.M. 2009. Evolution of glyphosate-resistant crop technology. *Weed Science* 57(1):108–117.

Green, J.M., C.B. Hazel, D.R. Forney, and L.M. Pugh. 2008. New multiple-herbicide crop resistance and formulation technology to augment the utility of glyphosate. *Pest Management Science* 64(4):332–339.

Greene, J.K., S.G. Turnipseed, M.J. Sullivan, and O.L. May. 2001. Treatment thresholds for stink bugs (Hemiptera: Pentatomidae) in cotton. *Journal of Economic Entomology* 94(2):403–409.

Greene, J.K., S. Bundy, P.M. Roberts, and B.R. Leonard. 2006. Identification and management of common boll-feeding bugs in cotton. August. EB-158. Clemson University. Blackville, SC. Available online at http://www.clemson.edu/psapublishing/PAGES/ENTOM/EB158.pdf. Accessed February 26, 2009.

Gressel, J. 1996. Fewer constraints than proclaimed to the evolution of glyphosate-resistant weeds. *Resistant Pest Management* 8(2):2–5.

Gressel, J., and L.A. Segel. 1990. Modeling the effectiveness of herbicide rotations and mixtures as strategies to delay or preclude resistance. *Weed Technology* 4:186–198.

Griffiths, B.S., S. Caul, J. Thompson, A.N.E. Birch, C. Scrimgeour, M.N. Andersen, J. Cortet, A. Messéan, C. Sausse, B. Lacroix, and P.H. Krogh. 2005. A comparison of soil microbial community structure, protozoa and nematodes in field plots of conventional and genetically modified maize expressing the *Bacillus thuringiensis* Cry1Ab toxin. *Plant and Soil* 275(1–2):135–146.

Griffiths, B.S., S. Caul, J. Thompson, A.N.E. Birch, C. Scrimgeour, J. Cortet, A. Foggo, C.A. Hackett, and P.H. Krogh. 2006. Soil microbial and faunal community responses to Bt maize and insecticide in two soils. *Journal of Environmental Quality* 35(3):734–741.

Gustafson, D.I. 2008. Sustainable use of glyphosate in North American cropping systems. *Pest Management Science* 64(4):409–416.

Hanson, B.D., C.A. Mallory-Smith, W.J. Price, B. Shafii, D.C. Thill, and R.S. Zemetra. 2005. Interspecific hybridization: Potential for movement of herbicide resistance from wheat to jointed goatgrass (*Aegilops cylindrica*). *Weed Technology* 19(3):674–682.

Hanson, J.D., J. Hendrickson, and D. Archer. 2008. Challenges for maintaining sustainable agricultural systems in the United States. *Renewable Agriculture and Food Systems* 23(4):325–334.

Hawes, C., A.J. Haughton, J.L. Osborne, D.B. Roy, S.J. Clark, J.N. Perry, P. Rothery, D.A. Bohan, D.R. Brooks, G.T. Champion, A.M. Dewar, M.S. Heard, I.P. Woiwod, R.E. Daniels, M.W. Young, A.M. Parish, R.J. Scott, L.G. Firbank, and G.R. Squire. 2003. Responses of plants and invertebrate trophic groups to contrasting herbicide regimes in the farm scale evaluations of genetically modified herbicide-tolerant crops. *Philosophical Transactions of the Royal Society B: Biological Sciences* 358(1439):1899–1913.

Hawes, C., A.J. Haughton, D.A. Bohan, and G.R. Squire. 2009. Functional approaches for assessing plant and invertebrate abundance patterns in arable systems. *Basic and Applied Ecology* 10(1):34–42.

Head, G., W. Moar, M. Eubanks, B. Freeman, J. Ruberson, A. Hagerty, and S. Turnipseed. 2005. A multiyear, large-scale comparison of arthropod populations on commercially managed Bt and non-Bt cotton fields. *Environmental Entomology* 34(5):1257–1266.

Heap, I. 2010. The international survey of herbicide resistant weeds. Available online at http://www.weedscience.org/. Accessed March 31, 2010.

Hernández-Martínez, P., J. Ferré, and B. Escriche. 2009. Broad-spectrum cross-resistance in *Spodoptera exigua* from selection with a marginally toxic Cry protein. *Pest Management Science* 65(6):645–650.

Heuberger, S., and Y. Carrière. 2009. Pollen-mediated transgene flow in agricultural seed production. Paper presented at the 94th Ecological Society of America annual meeting (Albuquerque, NM, August 2–7, 2009).

Heuberger, S., C. Ellers-Kirk, C. Yafuso, A.J. Gassmann, B.E. Tabashnik, T.J. Dennehy, and Y. Carrière. 2009. Effects of refuge contamination by transgenes on Bt resistance in pink bollworm (Lepidoptera: Gelechiidae). *Journal of Economic Entomology* 101(2):504–514.

Hilgenfeld, K.L., A.R. Martin, D.A. Mortensen, and S.C. Mason. 2004. Weed management in a glyphosate resistant soybean system: Weed species shifts. *Weed Technology* 18(2):284–291.

Holland, J.M. 2004. The environmental consequences of adopting conservation tillage in Europe: Reviewing the evidence. *Agriculture, Ecosystems and Environment* 103(1):1–25.

Hopkins, D.W., and E.G. Gregorich. 2003. Detection and decay of the Bt endotoxin in soil from a field trial with genetically modified maize. *European Journal of Soil Science* 54(4):793–800.

Howatt, K.A. 2008. Personal communication to M.D.K. Owen.

Howatt, K.A., G.J. Endres, P.E. Hendrickson, E.Z. Aberle, J.R. Lukach, B.M. Jenks, N.R. Riveland, S.A. Valenti, and C.M. Rystedt. 2006. Evaluation of glyphosate-resistant hard red spring wheat (*Triticum aestivum*). *Weed Technology* 20:706–716.

Hurley, T.M., and P.D. Mitchell. 2008. Insect resistance management: Adoption and compliance. In *Insect resistance management: Biology, economics, and prediction*. ed. D.W. Onstad, pp. 227–253. San Diego, CA: Elselvier.

Hutchison, W.D., E.C. Burkness, R.D. Moon, T. Leslie, S.J. Fleischer, and M. Abrahamson. 2007. Evidence for regional suppression of European corn borer populations in transgenic Bt maize in the Midwestern USA: Analysis of long-term time series data from three states, Vol. 2. In *Proceedings of the XVI International Plant Protection Congress*, pp. 512-513 (Glasgow, UK, October 15–18, 2007). British Crop Production Council.

Icoz, I., and G. Stotzky. 2008. Fate and effects of insect-resistant Bt crops in soil ecosystems. *Soil Biology and Biochemistry* 40(3):559–586.

Icoz, I., D. Saxena, D.A. Andow, C. Zwahlen, and G. Stotzky. 2008. Microbial populations and enzyme activities in soil in situ under transgenic corn expressing cry proteins from *Bacillus thuringiensis*. *Journal of Environmental Quality* 37(2):647–662.

Jackson, R.E., and H.N. Pitre. 2004. Influence of Roundup Ready® soybean production systems and glyphosate application on pest and beneficial insects in wide-row soybean. *Journal of Agricultural and Urban Entomology* 21(2):61–70.

Jackson, R.E., J.R. Bradley Jr., and J.W. Van Duyn. 2003. Field performance of transgenic cottons expressing one or two *Bacillus thuringiensis* endotoxins against bollworm, *Helicoverpa zea* (Boddie). *Journal of Cotton Science* 7(3):57–64.

Jaffe, G. 2009. Complacency on the farm: Significant noncompliance with EPA's refuge requirements threatens the future effectiveness of genetically engineered pest-protected corn. November. Center for Science in the Public Interest. Washington, DC. Available online at http://cspinet.org/new/pdf/complacencyonthefarm.pdf. Accessed August 5, 2009.

Jasieniuk, M., A.L. Brûlé-Babel, and I.N. Morrison. 1996. The evolution and genetics of herbicide resistance in weeds. *Weed Science* 44(1):176–193

Johnson, W.G., and K.D. Gibson. 2006. Glyphosate-resistant weeds and resistance management strategies: An Indiana grower perspective. *Weed Technology* 20(3):768–772.

Johnson, W.G., M.D.K. Owen, G.R. Kruger, B.G. Young, D.R. Shaw, R.G. Wilson, J.W. Wilcut, D.L. Jordan, and S.C. Weller. 2009. U.S. farmer awareness of glyphosate-resistant weeds and resistance management strategies. *Weed Technology* 23(2):308–312.

Jung, H.G., and C.C. Sheaffer. 2004. Influence of Bt transgenes on cell wall lignification and digestibility of maize stover for silage. *Crop Science* 44(5):1781–1789.

Kalaitzandonakes, N., and P. Suntornpithug. 2003. Adoption of cotton biotechnology in the United States: Implications for impact assessment. In *The economic and environmental impacts of agbiotech: A global perspective*. ed. N.G. Kalaitzandonakes, pp. 103–118. New York: Kluwer Academic/Plenum Publishers.

Keeler, K.H., C.E. Turner, and M.R. Bolick. 1996. Movement of crop transgenes into wild plants. In *Herbicide-resistant crops: Agricultural, environmental, economic, regulatory, and technical aspects*. ed. S.O. Duke, pp. 303–330. Boca Raton: Lewis Publishers.

Kennedy, G.G., and N.P. Storer. 2000. Life systems of polyphagous arthropod pests in temporally unstable cropping systems. *Annual Review of Entomology* 45:467–493.

Kennedy, G.G., F. Gould, O.M.B. Deponti, and R.E. Stinner. 1987. Ecological, agricultural, genetic, and commercial considerations in the deployment of insect-resistant germplasm. *Environmental Entomology* 16(2):327–338.

Kirkpatrick, K.J., and H.D. Wilson. 1988. Interspecific gene flow in *Cucurbita*: *C. texana* vs. *C. pepo*. *American Journal of Botany* 75(4):519–527.

Kladivko, E.J. 2001. Tillage systems and soil ecology. *Soil and Tillage Research* 61(1-2):61–76.

Knispel, A.L., S.M. McLachlan, R.C. Van Acker, and L.F. Friesen. 2008. Gene flow and multiple herbicide resistance in escaped canola populations. *Weed Science* 56(1):72–80.

Kniss, A.R., S.D. Miller, and R.G. Wilson. 2004. Factors influencing common lambsquarters control with glyphosate. In *Proceedings of the 59th Annual Meeting of the North Central Weed Science Society*, p. 85 (Columbus, OH, December 13–16, 2004). ed. R.G. Hartzler. North Central Weed Science Society. Available online at http://www.ncwss.org/proceed/2004/proc04/abstracts/085.pdf. Accessed January 31, 2009.

———. 2005. Common lambsquarters control with glyphosate: What's the problem? In *Proceedings of the 60th Annual Meeting of the North Central Weed Science Society*, p. 200 (Kansas City, MO, December 12–15, 2005). ed. R.G. Hartzler. North Central Weed Science Society. Available online at http://www.ncwss.org/proceed/2005/proc05/abstracts/200.pdf. Accessed July 24, 2009.

Kniss, A.R., D.J. Lyon, and S.D. Miller. 2008. Jointed goatgrass management with imazamox-resistant cultivars in a winter wheat-fallow rotation. *Crop Science* 48(6):2414–2420.

Kolpin, D.W., E.M. Thurman, E.A. Lee, M.T. Meyer, E.T. Furlong, and S.T. Glassmeyer. 2006. Urban contributions of glyphosate and its degradate AMPA to streams in the United States. *Science of the Total Environment* 354(2-3):191–197.

Kravchenko, A.N., X. Hao, and G.P. Robertson. 2009. Seven years of continuously planted Bt corn did not affect mineralizable and total soil C and total N in surface soil. *Plant and Soil* 318(1–2):269–274.

Kremer, R.J., and N.E. Means. 2009. Glyphosate and glyphosate-resistant crop interactions with rhizosphere microorganisms. *European Journal of Agronomy* 31(3):153–161.

Krogh, P.H., B. Griffiths, D. Demšar, M. Bohanec, M. Debeljak, M.N. Andersen, C. Sausse, A.N.E. Birch, S. Caul, M. Holmstrup, L.H. Heckmann, and J. Cortet. 2007. Responses by earthworms to reduced tillage in herbicide tolerant maize and Bt maize cropping systems. *Pedobiologia* 51(3):219–227.

Kruger, M., J.B.J. Van Rensburg, and J. Van den Berg. 2009. Perspective on the development of stem borer resistance to Bt maize and refuge compliance at the Vaalharts irrigation scheme in South Africa. *Crop Protection* 28(8):684–689.

Krupke, C., P. Marquardt, W. Johnson, S. Weller, and S.P. Conley. 2009. Volunteer corn presents new challenges for insect resistance management. *Agronomy Journal* 101(4):797–799.

Lachnicht, S.L., P.F. Hendrix, R.L. Potter, D.C. Coleman, and D.A. Crossley Jr. 2004. Winter decomposition of transgenic cotton residue in conventional-till and no-till systems. *Applied Soil Ecology* 27(2):135–142.

Landis, D.A., M.M. Gardiner, W. Van Der Werf, and S.M. Swinton. 2008. Increasing corn for biofuel production reduces biocontrol services in agricultural landscapes. *Proceedings of the National Academy of Sciences of the United States of America* 105(51):20552–20557.

Lang, A., R. Beck, J. Bauchhenß, G. Pommer, and M. Arndt. 2006. Monitoring of the environmental effects of the Bt gene. 10/2006. Bavarian State Research Center for Agriculture. August ed. Freising, Germany. Available online at http://www.lfl-neu.bayern.de/publikationen/daten/schriftenreihe_url_1_43.pdf. Accessed April 12, 2009.

Laughlin, K.D., A.G. Power, A.A. Snow, and L.J. Spencer. 2009. Risk assessment of genetically engineered crops: Fitness effects of virus-resistance transgenes in wild *Cucurbita pepo*. *Ecological Applications* 19(5):1091–1101.

Leer, S. 2006. Glyphosate-resistant giant ragweed confirmed in Indiana, Ohio. *Purdue University Newsletter*, December 21. Available online at http://www.agriculture.purdue.edu/agcomm/agnews/public/story.asp?newsid=1594. Accessed April 12, 2009.

Légère, A. 2005. Risks and consequences of gene flow from herbicide-resistant crops: Canola (*Brassica napus* L) as a case study. *Pest Management Science* 61(3):292–300.

Legleiter, T.R., and K.W. Bradley. 2008. Glyphosate and multiple herbicide resistance in common waterhemp (*Amaranthus rudis*) populations from Missouri. *Weed Science* 56(4):582–587.

Lehman, R.M., S.L. Osborne, and K.A. Rosentrater. 2008. No differences in decomposition rates observed between *Bacillus thuringiensis* and non-*Bacillus thuringiensis* corn residue incubated in the field. *Agronomy Journal* 100(1):163–168.

Liebig, M.A., D.L. Tanaka, and B.J. Wienhold. 2004. Tillage and cropping effects on soil quality indicators in the northern Great Plains. *Soil and Tillage Research* 78(2):131–141.

Liphadzi, K.B., K. Al-Khatib, C.N. Bensch, P.W. Stahlman, J.A. Dille, T. Todd, C.W. Rice, M.J. Horak, and G. Head. 2005. Soil microbial and nematode communities as affected by glyphosate and tillage practices in a glyphosate-resistant cropping system. *Weed Science* 53(4):536–545.

Liu, Y.B., B.E. Tabashnik, and M. Pusztai-Carey. 1996. Field-evolved resistance to *Bacillus thuringiensis* toxin Cry1C in diamondback moth (Lepidoptera: Plutellidae). *Journal of Economic Entomology* 89(4):798–804.

Llewellyn, D., C. Tyson, G. Constable, B. Duggan, S. Beale, and P. Steel. 2007. Containment of regulated genetically modified cotton in the field. *Agriculture, Ecosystems and Environment* 121(4):419–429.

Locke, M.A., and C.T. Bryson. 1997. Herbicide-soil interactions in reduced tillage and plant residue management systems. *Weed Science* 45(2):307–320.

Locke, M.A., R.M. Zablotowicz, and K.N. Reddy. 2008. Integrating soil conservation practices and glyphosate-resistant crops: Impacts on soil. *Pest Management Science* 64(4):457–469.

Loureiro, I., C. Escorial, J.-M. Garcia Baudin, and M.-C. Chueca. 2008. Hybridisation between wheat and *Aegilops geniculata* and hybrid fertility for potential herbicide resistance transfer. *Weed Research* 48:561–570.

Luna V., S., J. Figueroa M., B. Baltazar M., R. Gomez L., R. Townsend, and J.B. Schoper. 2001. Maize pollen longevity and distance isolation requirements for effective pollen control. *Crop Science* 41(5):1551–1557.

Mahon, R.J., and K.M. Olsen. 2009. Limited survival of a Cry2Ab-resistant strain of *Helicoverpa armigera* (Lepidoptera: Noctuidae) on Bollgard II. *Journal of Economic Entomology* 102(2):708–716.

Mahon, R.J., K.M. Olsen, S. Downes, and S. Addison. 2007. Frequency of alleles conferring resistance to the Bt toxins Cry1Ac and Cry2Ab in Australian populations of *Helicoverpa armigera* (Lepidoptera: Noctuidae). *Journal of Economic Entomology* 100(6):1844–1853.

Malik, J., G. Barry, and G. Kishore. 1989. The herbicide glyphosate. *Biofactors* 2(1):17–25.

Mallory-Smith, C., M. Butler, and C. Campbell. 2005. Gene movement from glyphosate-resistant creeping bentgrass (*Agrostis stolonifera*) fields, Abstracts, Vol. 45. In *Proceedings of the Weed Science Society of America*, pp. 49–50 (Honolulu, HI).

Manachini, B. 2003. Effect of transgenic corn on *Lydella thompsoni* Hertig (Diptera: Tachinidae) parasitoid of *Ostrinia nubilalis* Hb. (Lepidoptera: Crambidae). *Bollettino di Zoologia Agraria e di Bachicoltura* 35(111–125).

Marshall, M.W., K. Al-Khatib, and T. Loughin. 2001. Gene flow, growth, and competitiveness of imazethapyr-resistant common sunflower. *Weed Science* 49:14–21.

Marvier, M., C. McCreedy, J. Regetz, and P. Kareiva. 2007. A meta-analysis of effects of Bt cotton and maize on nontarget invertebrates. *Science* 316(5830):1475–1477.

Marvier, M., Y. Carrière, N. Ellstrand, P. Gepts, P. Kareiva, E. Rosi-Marshall, B.E. Tabashnik, and L.L. Wolfenbarger. 2008. Harvesting data from genetically engineered crops. *Science* 320(5875):452–453.

Massinga, R.A., K. Al-Khatib, P. St. Amand, and J.F. Miller. 2005. Relative fitness of imazamox-resistant common sunflower and prairie sunflower. *Weed Science* 53(2):166–174.

Massinga, R.A., and K. Al-Khatib. 2002. Gene flow of imidazolinone resistance from cultivated sunflower to common sunflower (*Helianthus annuus*) and prairie sunflower (*H. petiolaris*). In *Proceedings of the Weed Science Society of America*, Vol. 42. p. 21 (Reno, NV). ed. J. Wilcut. Weed Science Society of America.

Massinga, R.A., K. Al-Khatib, P. St. Amand, and J. F. Miller. 2003. Gene flow from imidazolinone-resistant domesticated sunflower to wild relatives. *Weed Science* 51:854–862.

Matten, S.R., G.P. Head, and H.D. Quemada. 2008. How governmental regulation can help or hinder the integration of Bt crops into IPM programs. In *Integration of insect-resistant genetically modified crops within IPM programs, Progress in biological control*. Vol. 5. eds. J. Romeis, A.M. Shelton, and G.G. Kennedy, pp. 27–39. New York: Springer.

Matus-Cádiz, M.A., P. Hucl, M.J. Horak, and L.K. Blomquist. 2004. Gene flow in wheat at the field scale. *Crop Science* 44(3):718–727.

Maxwell, B., and M. Jasieniuk. 2000. The evolution of herbicide resistance evolution models. In *Proceedings of the 3rd International Weed Science Congress*, p. 172 (Foz do Iguassu, Brazil, June 6–11, 2000). International Weed Science Society. Available online at http://www.iwss.info/docs/IWSC_2000.pdf. Accessed August 9, 2009.

McGregor, S.E. 1976. Insect pollination of cultivated crop plants. U.S. Department of Agriculture–Agricultural Research Service. Tucson, AZ. Available online at http://gears.tucson.ars.ag.gov/book/. Accessed July 1, 2009.

McKenzie, J.A. 1996. *Ecological and evolutionary aspects of insecticide resistance*. Austin, TX: R.G. Landes and Academic Press.

Men, X., F. Ge, C.A. Edwards, and E.N. Yardim. 2005. The influence of pesticide applications on *Helicoverpa armigera* Hübner and sucking pests in transgenic Bt cotton and non-transgenic cotton in China. *Crop Protection* 24(4):319–324.

Mendelsohn, M., J. Kough, Z. Vaituzis, and K. Matthews. 2003. Are Bt crops safe? *Nature Biotechnology* 21(9):1003–1009.

Mensah, E.C. 2007. *Economics of technology adoption: A simple approach*. Saarbrücken, Germany: VDM Verlag.

Micinski, S., D.C. Blouin, W.F. Waltman, and C. Cookson. 2008. Abundance of *Helicoverpa zea* and *Heliothis virescens* in pheromone traps during the past twenty years in Northwestern Louisiana. *Southwestern Entomologist* 33(2):139–149.

Mickelson, S.K., P. Boyd, J.L. Baker, and S.I. Ahmed. 2001. Tillage and herbicide incorporation effects on residue cover, runoff, erosion, and herbicide loss. *Soil and Tillage Research*
60(1–2):55–66.

Mitchell, P.D., and D.W. Onstad. 2005. Effect of extended diapause on evolution of resistance to transgenic *Bacillus thuringiensis* corn by northern corn rootworm (Coleoptera:
Chrysomelidae). *Journal of Economic Entomology* 98(6):2220–2234.

Moar, W.J., M. Pusztai-Carey, H.V. Faassen, D. Bosch, R. Frutos, C. Rang, K. Luo, and M.J.
Adang. 1995. Development of *Bacillus thuringiensis* Cry1C resistance by *Spodoptera
exigua* (Hubner) (Lepidoptera: Noctuidae). *Applied and Environmental Microbiology*
61(6):2086–2092.

Moar, W., R. Roush, A. Shelton, J. Ferré, S. MacIntosh, B.R. Leonard, and C. Abel. 2008. Field-
evolved resistance to Bt toxins. *Nature Biotechnology* 26(10):1072–1074.

Morin, S., R.W. Biggs, M.S. Sisterson, L. Shriver, C. Ellers-Kirk, D. Higginson, D. Holley,
L.J. Gahan, D.G. Heckel, Y. Carrière, T.J. Dennehy, J.K. Brown, and B.E. Tabashnik.
2003. Three cadherin alleles associated with resistance to *Bacillus thuringiensis* in pink
bollworm. *Proceedings of the National Academy of Sciences of the United States of America*
100(9):5004–5009.

Morrison, L.A., C.C. Lisele, and C.A. Mallory-Smith. 2002a. Infestations of jointed goatgrass
(*Aegilops cylindrica*) and its hybrids with wheat in Oregon wheat fields. *Weed Science*
50:737–747.

Morrison, L.A., O. Riera-Lizarazu, L. Cremieux, and C.A. Mallory-Smith. 2002b. Jointed
goatgrass (*Aegilops cylindrica* Host) X wheat (*Triticum aestivum* L.) hybrids: Hybridization dynamics in Oregon wheat fields. *Crop Science* 42:1863–1872.

Mueller, T.C., P.D. Mitchell, B.G. Young, and A.S. Culpepper. 2005. Proactive versus reactive
management of glyphosate-resistant or -tolerant weeds. *Weed Technology* 19(4):924–933.

Murphy, C.E., and D. Lemerle. 2006. Continuous cropping systems and weed selection.
Euphytica 148(1–2):61–73.

NAPPO (North American Plant Protection Organization). 2003. *Aegilops cylindrica* Host. PRA/
Grains panel pest facts sheet. North American Plant Protection Organization. Ottawa,
Ontario. Available online at http://www.nappo.org/PRA-sheets/Aegilopscylindrica.
pdf. Accessed November 6, 2009.

Naranjo, S.E. 2005. Long-term assessment of the effects of transgenic Bt cotton on the function of the natural enemy community. *Environmental Entomology* 34(5):1211–1223.

———. 2009. Impacts of Bt crops on non-target invertebrates and insecticide use patterns.
Perspectives in Agriculture, Veterinary Science, Nutrition and Natural Resources 4(11):1–11.

Neve, P. 2008. Simulation modelling to understand the evolution and management of
glyphosate resistance in weeds. *Pest Management Science* 64(4):392–401.

Neve, P., A.J. Diggle, F.P. Smith, and S.B. Powles. 2003. Simulating evolution of glyphosate
resistance in *Lolium rigidum* I: Population biology of a rare resistance trait. *Weed Research*
43(6):404–417.

NRC (National Research Council). 2002. *Environmental effects of transgenic plants: The scope
and adequacy of regulation*. Washington, DC: National Academy Press.

Oliver, L.R., S.A. Harrison, and M. McClelland. 1983. Germination of Texas gourd (*Cucurbita
texana*) and its control in soybeans (*Glycine max*). *Weed Science* 31(5):700–706.

Olsen, K.M., J.C. Daly, H.E. Holt, and E.J. Finnegan. 2005. Season-long variation in expression of Cry1Ac gene and efficacy of *Bacillus thuringiensis* toxin in transgenic cotton
against *Helicoverpa armigera* (Lepidoptera: Noctuidae). *Journal of Economic Entomology*
98(3):1007–1017.

Onstad, D.W. 2008. *Insect resistance management: Biology, economics, and prediction*. San Diego,
CA: Elselvier.

Orr, D.B., and D.A. Landis. 1997. Oviposition of European corn borer (Lepidoptera: Pyralidae) and impact of natural enemy populations in transgenic versus isogenic corn. *Journal of Economic Entomology* 90(4):905–909.

Owen, M.D.K. 2001. Importance of weed population shifts and herbicide resistance in the Midwest USA Corn Belt. In *Pesticide behaviour in soils and water: Proceedings of a symposium organised by the British Crop Protection Council*, pp. 407–412 (Brighton, UK, November 13–15, 2001). Farmham, Surrey, UK: British Crop Protection Council.

———. 2008. Weed species shifts in glyphosate-resistant crops. *Pest Management Science* 64(4):377–387.

Owen, M.D.K., and I.A. Zelaya. 2005. Herbicide-resistant crops and weed resistance to herbicides. *Pest Management Science* 61(3):301–311.

Ozinga, W.A., R.M. Bekker, J.H.J. Schaminée, and J.M. Van Groenendael. 2004. Dispersal potential in plant communities depends on environmental conditions. *Journal of Ecology* 92(5):767–777.

Palmer, R.G., J. Gai, H. Sun, and J.W. Burton. 2001. Production and evaluation of hybrid soybean. In *Plant breeding reviews*, Vol. 20. ed. J. Janick, pp. 263-308. New York: John Wiley & Sons, Inc.

Palumbi, S.R. 2001. Humans as the world's greatest evolutionary force. *Science* 293(5536):1786–1790.

Patzoldt, W.L., P.J. Tranel, and A.G. Hager. 2005. A waterhemp (*Amaranthus tuberculatus*) biotype with multiple resistance across three herbicide sites of action. *Weed Science* 53(1):30–36.

Pereira, E.J.G., N.P. Storer, and B.D. Siegfried. 2008. Inheritance of Cry1F resistance in laboratory-selected European corn borer and its survival on transgenic corn expressing the Cry1F toxin. *Bulletin of Entomological Research* 98(6):621–629.

Peterson, R.K.D., and A.G. Hulting. 2004. A comparative ecological risk assessment for herbicides used on spring wheat: The effect of glyphosate when used within a glyphosate-tolerant wheat system. *Weed Science* 52(5):834–844.

Pietrantonio, P.V., T.A. Junek, R. Parker, D. Mott, K. Siders, N. Troxclair, J. Vargas-Camplis, J.K. Westbrook, and V.A. Vassiliou. 2007. Detection and evolution of resistance to the pyrethroid cypermethrin in *Helicoverpa zea* (Lepidoptera: Noctuidae) populations in Texas. *Environmental Entomology* 36(5):1174–1188.

Poerschmann, J., A. Gathmann, J. Augustin, U. Langer, and T. Górecki. 2005. Molecular composition of leaves and stems of genetically modified Bt and near-isogenic non-Bt maize—characterization of lignin patterns. *Journal of Environmental Quality* 34(5):1508–1518.

Pons, X., and P. Starý. 2003. Spring aphid-parasitoid (Hom., Aphididae, Hym., Braconidae) associations and interactions in a Mediterranean arable crop ecosystem, including Bt maize. *Anzeiger fur Schadlingskunde* 76(5):133–138.

Powles, S.B. 2008. Evolved glyphosate-resistant weeds around the world: Lessons to be learnt. *Pest Management Science* 64(4):360–365.

Powles, S.B., and C. Preston. 2006. Evolved glyphosate resistance in plants: Biochemical and genetic basis of resistance. *Weed Technology* 20(2):282–289.

Powles, S.B., D.F. Lorraine-Colwill, J.J. Dellow, and C. Preston. 1998. Evolved resistance to glyphosate in rigid ryegrass (*Lolium rigidum*) in Australia. *Weed Science* 46(5):604–607.

Puricelli, E., and D. Tuesca. 2005. Weed density and diversity under glyphosate-resistant crop sequences. *Crop Protection* 24(6):533–542.

Quemada, H., L. Strehlow, D. Decker-Walters, S., and J. Staub, E. 2008. Population size and incidence of virus infection in free-living populations of *Cucurbita pepo*. *Environmental Biosafety Research* 7(4):185–196.

Rehman, M., J.L. Hansen, J. Brown, W. Price, R.S. Zemetra, and C.A. Mallory-Smith. 2006. Effect of wheat genotype on the phenotype of wheat x jointed goatgrass (*Aegilops cylindrica*) hybrids. *Weed Science* 54(4):690–694.

Reichman, J.R., L.S. Watrud, E.H. Lee, C.A. Burdick, M.A. Bollman, M.J. Storm, G.A. King, and C. Mallory-Smith. 2006. Establishment of transgenic herbicide-resistant creeping bentgrass (*Agrostis stolonifera* L.) in nonagronomic habitats. *Molecular Ecology* 15(13):4243–4255.

Relyea, R.A., and D.K. Jones. 2009. The toxicity of Roundup Original Max® to 13 species of larval amphibians. *Environmental Toxicology and Chemistry* 28(9):2004–2008.

Riggin-Bucci, T.M., and F. Gould. 1997. Impact of intraplot mixtures of toxic and nontoxic plants on population dynamics of diamondback moth (Lepidoptera: Plutellidae) and its natural enemies. *Journal of Economic Entomology* 90(2):241–251.

Roberts, R.K., B.C. English, Q. Gao, and J.A. Larson. 2006. Simultaneous adoption of herbicide-resistance and conservation-tillage cotton technologies. *Journal of Agricultural and Applied Economics* 38(3):629–643.

Romeis, J., M. Meissle, and F. Bigler. 2006. Transgenic crops expressing *Bacillus thuringiensis* toxins and biological control. *Nature Biotechnology* 24(1):63–71.

Rosi-Marshall, E.J., J.L. Tank, T.V. Royer, M.R. Whiles, M. Evans-White, C. Chambers, N.A. Griffiths, J. Pokelsek, and M.L. Stephen. 2007. Toxins in transgenic crop byproducts may affect headwater stream ecosystems. *Proceedings of the National Academy of Sciences of the United States of America* 104(41):16204–16208.

Roush, R. 1997. Managing resistance to transgenic crops. In *Advances in insect control: The role of transgenic plants*. eds. N. Carozzi and M. Koziel, pp. 271–294. London: Taylor & Francis Ltd.

Roush, R.T. 1998. Two-toxin strategies for management of insecticidal transgenic crops: Can pyramiding succeed where pesticide mixtures have not? *Philosophical Transactions of the Royal Society B: Biological Sciences* 353(1376):1777–1786.

Roux, F., M. Paris, and X. Reboud. 2008. Delaying weed adaptation to herbicide by environmental heterogeneity: A simulation approach. *Pest Management Science* 64(1):16–29.

Sammons, R.D., D.C. Heering, N. Dinicola, H. Glick, and G.A. Elmore. 2007. Sustainability and stewardship of glyphosate and glyphosate-resistant crops. *Weed Technology* 21(2):347–354.

Sasu, M.A., M.J. Ferrari, D. Du, J.A. Winsor, and A.G. Stephenson. 2009. Indirect costs of a nontarget pathogen mitigate the direct benefits of a virus-resistant transgene in wild *Cucurbita*. *Proceedings of the National Academy of Sciences of the United States of America* 106(45):19067–19071.

Saxena, D., and G. Stotzky. 2001a. *Bacillus thuringiensis* (Bt) toxin released from root exudates and biomass of Bt corn has no apparent effect on earthworms, nematodes, protozoa, bacteria, and fungi in soil. *Soil Biology and Biochemistry* 33(9):1225–1230.

———. 2001b. Bt corn has a higher lignin content than non-Bt corn. *American Journal of Botany* 88(9):1704–1706.

Schmale, D., R. Anderson, D. Lyon, and B. Klein. 2008. Jointed goatgrass: Best management practices, central Great Plains. EB2033E. Washington State University. Pullman, WA. Available online at http://cru.cahe.wsu.edu/CEPublications/eb2033e/eb2033e.pdf. Accessed August 5, 2009.

Schmale, D., T. Peeper, and P. Stahlman. 2009. Jointed goatgrass: Best management practices, southern Great Plains. EM011. Washington State University. Pullman, WA. Available online at http://jointedgoatgrass.wsu.edu/jointedgoatgrass/bulletins/EM011_Final_Version.pdf. Accessed September 14, 2009.

Schnepf, E., N. Crickmore, J. Van Rie, D. Lereclus, J. Baum, J. Feitelson, D.R. Zeigler, and D.H. Dean. 1998. *Bacillus thuringiensis* and its pesticidal crystal proteins. *Microbiology and Molecular Biology Reviews* 62(3):775–806.

Schuler, T.H., R.P.J. Potting, I. Denholm, S.J. Clark, A.J. Clark, C.N. Stewart, and G.M. Poppy. 2003. Tritrophic choice experiments with Bt plants, the diamondback moth (*Plutella xylostella*) and the parasitoid *Cotesia plutellae*. *Transgenic Research* 12(3):351–361.

Schuler, T.H., I. Denholm, S.J. Clark, C.N. Stewart, and G.M. Poppy. 2004. Effects of Bt plants on the development and survival of the parasitoid *Cotesia plutellae* (Hymenoptera: Braconidae) in susceptible and Bt-resistant larvae of the diamondback moth, *Plutella xylostella* (Lepidoptera: Plutellidae). *Journal of Insect Physiology* 50(5):435–443.

Schuster, C.L., D.E. Shoup, and K. Al-Khatib. 2007. Response of common lambsquarters (*Chenopodium album*) to glyphosate as affected by growth stage. *Weed Science* 55(2):147–151.

Scott, B.A., and M.J. VanGessel. 2007. Delaware soybean grower survey on glyphosate-resistant horseweed (*Conyza canadensis*). *Weed Technology* 21(1):270–274.

Scursoni, J.A., F. Forcella, and J. Gunsolus. 2007. Weed escapes and delayed weed emergence in glyphosate-resistant soybean. *Crop Protection* 26(3):212–218.

Shaner, D.L. 2000. The impact of glyphosate-tolerant crops on the use of other herbicides and on resistance management. *Pest Management Science* 56(4):320–326.

Shelton, A.M., S.L. Hatch, J.Z. Zhao, M. Chen, E.D. Earle, and J. Cao. 2008. Suppression of diamondback moth using Bt-transgenic plants as a trap crop. *Crop Protection* 27(3–5):403–409.

Shields, E.J., J.T. Dauer, M.J. VanGessel, and G. Neumann. 2006. Horseweed (*Conyza canadensis*) seed collected in the planetary boundary layer. *Weed Science* 54(6):1063–1067.

Shipitalo, M.J., and L.B. Owens. 2006. Tillage system, application rate, and extreme event effects on herbicide losses in surface runoff. *Journal of Environmental Quality* 35(6):2186–2194.

Shipitalo, M.J., R.W. Malone, and L.B. Owens. 2008. Impact of glyphosate-tolerant soybean and glufosinate-tolerant corn production on herbicide losses in surface runoff. *Journal of Environmental Quality* 37(2):401–408.

Showalter, A.M., S. Heuberger, B.E. Tabashnik, and Y. Carrière. 2009. A primer for the use of insecticidal transgenic cotton in developing countries. *Journal of Insect Science* 9(22):1–39.

Siegfried, B.D., A.C. Zoerb, and T. Spencer. 2001. Development of European corn borer larvae on Event 176 Bt corn: Influence on survival and fitness. *Entomologia Experimentalis et Applicata* 100(1):15–20.

Simard, M.-J., A. Légère, and S.I. Warwick. 2006. Transgenic *Brassica napus* fields and *Brassica rapa* weeds in Quebec: Sympatry and weed-crop in situ hybridization. *Canadian Journal of Botany* 84(12):1842–1851.

Sisterson, M.S., L. Antilla, Y. Carrière, C. Ellers-Kirk, and B.E. Tabashnik. 2004a. Effects of insect population size on evolution of resistance to transgenic crops. *Journal of Economic Entomology* 97(4):1413–1424.

Sisterson, M.S., R.W. Biggs, C. Olson, Y. Carrière, T.J. Dennehy, and B.E. Tabashnik. 2004b. Arthropod abundance and diversity in Bt and non-Bt cotton fields. *Environmental Entomology* 33(4):921–929.

Sisterson, M.S., R.W. Biggs, N.M. Manhardt, Y. Carrière, T.J. Dennehy, and B.E. Tabashnik. 2007. Effects of transgenic Bt cotton on insecticide use and abundance of two generalist predators. *Entomologia Experimentalis et Applicata* 124(3):305–311.

Snow, A.A. 2002. Transgenic crops—why gene flow matters. *Nature Biotechnology* 20(6):542.

Snow, A.A., and P. Palma. 1997. Commercialization of transgenic plants: Potential ecological risks. *BioScience* 47(2):86–96.

Spencer, L.J., and A.A. Snow. 2001. Fecundity of transgenic wild-crop hybrids of *Cucurbita pepo* (Cucurbitaceae): Implications for crop-to-wild gene flow. *Heredity* 86(6):694–702.

Stewart, C.N., Jr., M.D. Halfhill, and S.I. Warwick. 2003. Transgene introgression from genetically modified crops to their wild relatives. *Nature Reviews Genetics* 4(10):806–817.

Storer, N.P., S.L. Peck, F. Gould, J.W. Van Duyn, and G.G. Kennedy. 2003. Spatial processes in the evolution of resistance in *Helicoverpa zea* (Lepidoptera: Noctuidae) to Bt transgenic corn and cotton in a mixed agroecosystem: A biology-rich stochastic simulation model. *Journal of Economic Entomology* 96(1):156–172.

Storer, N.P., G.P. Dively, and R.A. Herman. 2008. Landscape effects of insect-resistant genetically modified crops. In *Integration of insect-resistant genetically modified crops within IPM programs*. Vol. 5. eds. J. Romeis, A.M. Shelton, and G.G. Kennedy, pp. 273–302. New York: Springer Science + Business Media.

Stubbendieck, J.L., G.Y. Friisoe, and M.R. Bolick. 1994. *Weeds of Nebraska and the Great Plains*. Lincoln: Nebraska Department of Agriculture, Bureau of Plant Industry in cooperation with the University of Nebraska-Lincoln.

Tabashnik, B.E., and Y. Carrière. 2008. Evolution of insect resistance to transgenic plants. In *Specialization, speciation, and radiation: The evolutionary biology of herbivorous insects*. ed. K.J. Tilmon, pp. 267–279. Berkeley: University of California Press.

Tabashnik, B.E., and Y. Carrière. 2009. Insect resistant GM crops; pest resistance. In *Environmental impact of genetically modified crops*. eds. N. Ferry and A.M.R. Gatehouse, pp. 74–100. Wallingford, UK: CAB International.

Tabashnik, B.E., N. Finson, M.W. Johnson, and W.J. Moar. 1993. Resistance to toxins from *Bacillus thuringiensis* subsp. *kurstaki* causes minimal cross-resistance to *B. thuringiensis* subsp. *aizawai* in the diamondback moth (Lepidoptera: Plutellidae). *Applied and Environmental Microbiology* 59(5):1332–1335.

Tabashnik, B.E., A.L. Patin, T.J. Dennehy, Y.B. Liu, E. Miller, and R.T. Staten. 1999. Dispersal of pink bollworm (Lepidoptera: Gelechiidae) males in transgenic cotton that produces a *Bacillus thuringiensis* toxin. *Journal of Economic Entomology* 92(4):772–780.

Tabashnik, B.E., A.L. Patin, T.J. Dennehy, Y.B. Liu, Y. Carrière, M.A. Sims, and L. Antilla. 2000. Frequency of resistance to *Bacillus thuringiensis* in field populations of pink bollworm. *Proceedings of the National Academy of Sciences of the United States of America* 97(24):12980–12984.

Tabashnik, B.E., F. Gould, and Y. Carrière. 2004. Delaying evolution of insect resistance to transgenic crops by decreasing dominance and heritability. *Journal of Evolutionary Biology* 17(4):904–912.

Tabashnik, B.E., T.J. Dennehy, and Y. Carrière. 2005. Delayed resistance to transgenic cotton in pink bollworm. *Proceedings of the National Academy of Sciences of the United States of America* 102(43):15389–15393.

Tabashnik, B.E., J.A. Fabrick, S. Henderson, R.W. Biggs, C.M. Yafuso, M.E. Nyboer, N.M. Manhardt, L.A. Coughlin, J. Sollome, Y. Carrière, T.J. Dennehy, and S. Morin. 2006. DNA screening reveals pink bollworm resistance to Bt cotton remains rare after a decade of exposure. *Journal of Economic Entomology* 99(5):1525–1530.

Tabashnik, B.E., A.J. Gassmann, D.W. Crowder, and Y. Carrière. 2008. Insect resistance to Bt crops: Evidence versus theory. *Nature Biotechnology* 26(2):199–202.

———. 2008. Field-evolved resistance to Bt toxins. *Nature Biotechnology* 26(10):1074–1076.

Tabashnik, B.E., J.B.J. Van Rensburg, and Y. Carrière. 2009a. Field-evolved insect resistance to Bt crops: Definition, theory, and data. *Journal of Economic Entomology* 102:2011–2025.

Tabashnik, B.E., G.C. Unnithan, L. Masson, D.W. Crowder, X. Li, and Y. Carrière. 2009b. Asymmetrical cross-resistance between *Bacillus thuringiensis* toxins Cry1Ac and Cry2Ab in pink bollworm. *Proceedings of the National Academy of Sciences of the United States of America* 106(29):11889–11894.

Tarkalson, D.D., S.D. Kachman, J.M.N. Knops, J.E. Thies, and C.S. Wortmann. 2008. Decomposition of Bt and non-Bt corn hybrid residues in the field. *Nutrient Cycling in Agroecosystems* 80(3):211–222.

Terán-Vargas, A.P., J.C. Rodríguez, C.A. Blanco, J.L. Martínez-Carrillo, J. Cibrián-Tovar, H. Sánchez-Arroyo, L.A. Rodríguez-del-Bosque, and D. Stanley. 2005. Bollgard cotton and resistance of tobacco budworm (Lepidoptera: Noctuidae) to conventional insecticides in Southern Tamaulipas, Mexico. *Journal of Economic Entomology* 98(6):2203–2209.

Thompson, G.D., S. Matten, I. Denholm, M.E. Whalon, and P. Leonard. 2008. The politics of resistance management: Working towards pesticide resistance mangement globally. In *Global pesticide resistance in arthropods.* eds. M.E. Whalon, D. Mota-Sanchez and R.M. Hollingworth, pp. 146–164. Wallingford, UK: CAB International.

Tsui, M.T.K., and L.M. Chu. 2003. Aquatic toxicity of glyphosate-based formulations: Comparison between different organisms and the effects of environmental factors. *Chemosphere* 52(7):1189–1197.

Tuesca, D., E. Puricelli, and J.C. Papa. 2001. A long-term study of weed flora shifts in different tillage systems. *Weed Research* 41(4):369–382.

Uri, N.D., J.D. Atwood, and J. Sanabria. 1999. The environmental benefits and costs of conservation tillage. *Environmental Geology* 38(2):111–125.

US-EPA (U.S. Environmental Protection Agency). 1998. The Environmental Protection Agency's white paper on Bt plant-pesticide resistance management. January 14. Washington, DC. Available online at http://www.epa.gov/EPA-PEST/1998/January/Day-14/paper.pdf. Accessed May 4, 2009.

———. 2000. Great Lakes binational toxics strategy: The level I pesticides in the binational strategy. March 1. Section 3. Washington, DC. Available online at http://www.epa.gov/greatlakes/bns/pesticides/finalpestreport.html. Accessed April 13, 2009.

———. 2001. Biopesticides registration action document—*Bacillus thuringiensis* plant-incorporated protectants. October 16. Washington, DC. Available online at http://www.epa.gov/pesticides/biopesticides/pips/bt_brad.htm. Accessed June 13, 2009.

———. 2005. Biopesticides registration action document: *Bacillus thuringiensis* var. *aizawai* Cry1F and the genetic material (from the insert of plasmid PGMA281) necessary for its production in cotton and *Bacillus thuringiensis* var. *kurstaki* Cry1Ac and the genetic material (from the insert of plasmid PMYC3006) necessary for its production in cotton. Washington, DC. Available online at http://www.epa.gov/pesticides/biopesticides/ingredients/tech_docs/brad_006512-006513.pdf. Accessed April 20, 2009.

———. 2006a. A set of scientific issues being considered by the Environmental Protection Agency regarding: Evaluation of the resistance risks from using 100% Bollgard and Bollgard II cotton as part of a pink bollworm eradication program in the state of Arizona. October 24–26. SAP Minutes No 2007-01. Washington, DC. Available online at http://www.epa.gov/scipoly/sap/meetings/2006/october/october2006finalmeetingminutes.pdf. Accessed March 31, 2010.

———. 2006b. A set of scientific issues being considered by the Environmental Protection Agency regarding: Analysis of a natural refuge of non-cotton hosts for Monsanto's Bollgard II cotton. September 8. SAP Minutes No 2006-03. Washington, DC. Available online at http://www.epa.gov/scipoly/sap/meetings/2006/june/june2006finalmeetingminutes.pdf. Accessed January 22, 2009.

———. 2007. Pesticide news story: EPA approves natural refuge for insect resistance management in Bollgard II cotton. Washington, DC. Available online at http://www.epa.gov/oppfead1/cb/csb_page/updates/2007/bollgard-cotton.htm. Accessed January 22, 2009.

———. 2008a. Insect resistance management fact sheet for *Bacillus thuringiensis* (Bt) cotton products. Washington, DC. Available online at http://www.epa.gov/opp00001/biopesticides/pips/bt_cotton_refuge_2006.htm. Accessed January 22, 2009.

———. 2008b. Insect resistance management fact sheet for *Bacillus thuringiensis* (Bt) corn products. Washington, DC. Available online at http://www.epa.gov/oppbppd1/biopesticides/pips/bt_corn_refuge_2006.htm. Accessed January 22, 2009.

USDA-APHIS (U.S. Department of Agriculture–Animal and Plant Health Inspection Service). 2008. Determination of non-regulated status for glyphosate-tolerant (GlyTol™) cotton, *Gossypium hirsutum*, event GHB614. May 16. Petition 06-332-01p. Washington, DC. Available online at http://www.aphis.usda.gov/brs/aphisdocs/06_33201p_pea.pdf. Accessed March 31, 2010.

USDA-ERS (U.S. Department of Agriculture–Economic Research Service). 2009. Agricultural biotechnology: Adoption of biotechnology and its production impacts. Washington, DC. Available online at http://www.ers.usda.gov/Briefing/Biotechnology/chapter1.htm. Accessed October 1, 2009.

———. 2006. Agricultural resources and environmental indicators, 2006 edition. EIB-16. Washington, DC. Available online at http://www.ers.usda.gov/publications/arei/eib16/. Accessed February 17, 2009.

USDA-NASS (U.S. Department of Agriculture–National Agricultural Statistics Service). 2001. Acreage. June 29. Cr Pr 2-5 (6-01). Washington, DC. Available online at http://usda.mannlib.cornell.edu/usda/nass/Acre//2000s/2001/Acre-06-29-2001.pdf. Accessed April 14, 2009.

———. 2003. Acreage. June 30. Cr Pr 2-5 (6-03). Washington, DC. Available online at http://usda.mannlib.cornell.edu/usda/nass/Acre//2000s/2003/Acre-06-30-2003.pdf. Accessed April 14, 2009.

———. 2005. Acreage. June 30. Cr Pr 2-5 (6-05). Washington, DC. Available online at http://usda.mannlib.cornell.edu/usda/nass/Acre//2000s/2005/Acre-06-30-2005.pdf. Accessed April 14, 2009.

———. 2007. Acreage. June 29. Cr Pr 2-5 (6-07). Washington, DC. Available online at http://usda.mannlib.cornell.edu/usda/nass/Acre//2000s/2007/Acre-06-29-2007.pdf. Accessed April 14, 2009.

———. 2008. Acreage. June 30. Cr Pr 2-5 (6-08). Washington, DC. Available online at http://usda.mannlib.cornell.edu/usda/nass/Acre//2000s/2008/Acre-06-30-2008.pdf. Accessed March 31, 2010.

———. 2009a. Data and statistics: Quick stats. Washington, DC. Available online at http://www.nass.usda.gov/Data_and_Statistics/Quick_Stats/index.asp. Accessed June 22, 2009.

———. 2009b. Acreage. June 30. Cr Pr 2-5 (6-09).Washington, DC. Available online at http://usda.mannlib.cornell.edu/usda/current/Acre/Acre-06-30-2009.pdf. Accessed November 24, 2009.

Van Deynze, A.E., F.J. Sundstrom, and K.J. Bradford. 2005. Pollen-mediated gene flow in California cotton depends on pollinator activity. *Crop Science* 45(4):1565–1570.

van Frankenhuyzen, K., and C. Nystrom. 2002. The *Bacillus thuringiensis* toxin specificity database. The Great Lakes Forestry Centre. Available online at http://www.glfc.forestry.ca/bacillus/. Accessed April 27, 2009.

van Rensburg, J.B.J. 2007. First report of field resistance by the stem borer, *Busseola fusca* (Fuller) to Bt-transgenic maize. *South African Journal of Plant and Soil* 24(3):147–151.

VanGessel, M.J. 2001. Glyphosate-resistant horseweed from Delaware. *Weed Science* 49:703–705.

Vercesi, M.L., P.H. Krogh, and M. Holmstrup. 2006. Can *Bacillus thuringiensis* (Bt) corn residues and Bt-corn plants affect life-history traits in the earthworm *Aporrectodea caliginosa*? *Applied Soil Ecology* 32(2):180–187.

Walker, K., M. Mendelsohn, S. Matten, M. Alphin, and D. Ave. 2003a. The role of microbial Bt products in U.S. crop protection. *Journal of New Seeds* 5(1):31–51.

Walker, K.R., M.D. Ricciardone, and J. Jensen. 2003b. Developing an international consensus on DDT: A balance of environmental protection and disease control. *International Journal of Hygiene and Environmental Health* 206(4-5):423–435.

Walklate, P.J., J.C.R. Hunt, H.L. Higson, and J.B. Sweet. 2004. A model of pollen-mediated gene flow for oilseed rape. *Proceedings of the Royal Society B: Biological Sciences* 271(1538):441–449.

Wang, P., J.Z. Zhao, A. Rodrigo-Simón, W. Kain, A.F. Janmaat, A.M. Shelton, J. Ferré, and J. Myers. 2007. Mechanism of resistance to *Bacillus thuringiensis* toxin Cry1Ac in a greenhouse population of the cabbage looper, *Trichoplusia ni*. *Applied and Environmental Microbiology* 73(4):1199–1207.

Wardle, D.A. 1995. Impacts of disturbance on detritus food webs in agro-ecosystems of contrasting tillage and weed management practices. *Advances in Ecological Research* 26:105–185.

Warwick, S.I., M.J. Simard, A. Légère, H.J. Beckie, L. Braun, B. Zhu, P. Mason, G. Séguin-Swartz, and C.N. Stewart Jr. 2003. Hybridization between transgenic *Brassica napus* L. and its wild relatives: *Brassica rapa* L., *Raphanus raphanistrum* L., *Sinapis arvensis* L., and *Erucastrum gallicum* (Willd.) O.E. Schulz. *Theoretical and Applied Genetics* 107(3):528–539.

Warwick, S.I., A. Légère, M.J. Simard, and T. James. 2008. Do escaped transgenes persist in nature? The case of an herbicide resistance transgene in a weedy *Brassica rapa* population. *Molecular Ecology* 17(5):1387–1395.

Warwick, S.I., A. Francis, and J. La Fleche, eds. 2000. *Guide to wild germplasm of Brassica and allied crops (tribe Brassiceae, Brassicaceae).* 2nd ed. Agriculture and Agri-Food Canada. Ottawa, Ontario, Canada: AAFC-ECORC.

Wauchope, R.D., T.M. Buttler, A.G. Hornsby, P.W.M. Augustijn-Beckers, and J.P. Burt. 1992. The SCS/ARS/CES pesticide properties database for environmental decision-making. *Reviews of Environmental Contamination and Toxicology* 123:1–156.

Wauchope, R.D., T.L. Estes, R. Allen, J.L. Baker, A.G. Hornsby, R.L. Jones, R.P. Richards, and D.I. Gustafson. 2002. Predicted impact of transgenic, herbicide-tolerant corn on drinking water quality in vulnerable watersheds of the mid-western USA. *Pest Management Science* 58(2):146–160.

Weaver, M.A., L.J. Krutz, R.M. Zablotowicz, and K.N. Reddy. 2007. Effects of glyphosate on soil microbial communities and its mineralization in a Mississippi soil. *Pest Management Science* 63(4):388–393.

Weise, E. 2007. Effects of genetically engineered alfalfa cultivate a debate. *USA Today*, February 15. p. 10D, Life section. Available online at http://www.usatoday.com/news/health/2007-02-14-alfalfa_x.htm. Accessed June 4, 2009.

Werth, J.A., C. Preston, I.N. Taylor, G.W. Charles, G.N. Roberts, and J. Baker. 2008. Managing the risk of glyphosate resistance in Australian glyphosate-resistant cotton production systems. *Pest Management Science* 64(4):417–421.

Westgate, M.E., J. Lizaso, and W. Batchelor. 2003. Quantitative relationships between pollen shed density and grain yield in maize. *Crop Science* 43(3):934–942.

WHO (World Health Organization). 2005. Glyphosate and AMPA in drinking-water. Geneva: World Health Organization. Available online at http://www.who.int/water_sanitation_health/dwq/chemicals/glyphosate/en/. Accessed March 11, 2009.

Williams, M.R. 2009. Cotton insect losses—2008. Paper presented at the Beltwide Cotton Conference (San Antonio, TX, January 5–8, 2008). National Cotton Council of America. Available online at http://ncc.confex.com/ncc/2009/webprogram/Paper9858.html. Accessed March 25, 2009.

Wilson, R.G., S.D. Miller, P. Westra, A.R. Kniss, P.W. Stahlman, G.W. Wicks, and S.D. Kachman. 2007. Glyphosate-induced weed shifts in glyphosate-resistant corn or a rotation of glyphosate-resistant corn, sugarbeet, and spring wheat. *Weed Technology* 21(4):900–909.

Wolfenbarger, L.L., S. Naranjo, J. Lundgren, R. Bitzer, and L. Watrud. 2008. Bt crop effects on functional guilds of non-target arthropods: A meta-analysis. *PLoS ONE* 3(5).

Wu, K. 2007. Monitoring and management strategy for *Helicoverpa armigera* resistance to Bt cotton in China. *Journal of Invertebrate Pathology* 95(3):220–223.

Wu, K., W. Li, H. Feng, and Y. Guo. 2002. Seasonal abundance of the mirids, *Lygus lucorum* and *Adelphocoris* spp. (Hemiptera: Miridae) on Bt cotton in northern China. *Crop Protection* 21(10):997–1002.

Wu, K., W. Mu, G. Liang, and Y. Guo. 2005. Regional reversion of insecticide resistance in *Helicoverpa armigera* (Lepidoptera: Noctuidae) is associated with the use of Bt cotton in northern China. *Pest Management Science* 61(5):491–498.

Wu, K.-M., Y.-H. Lu, H.-Q. Feng, Y.-Y. Jiang, and J.-Z. Zhao. 2008. Suppression of cotton bollworm in multiple crops in China in areas with Bt toxin-containing cotton. *Science* 321(5896):1676–1678.

Xanthopoulos, F.P., and U.E. Kechagia. 2000. Natural crossing in cotton (*Gossypium hirsutum* L.). *Australian Journal of Agricultural Research* 51(8):979–983.

Yang, Y., H. Chen, Y. Wu, Y. Yang, and S. Wu. 2007. Mutated cadherin alleles from a field population of *Helicoverpa armigera* confer resistance to *Bacillus thuringiensis* toxin Cry1Ac. *Applied Environmental Microbiology* 73(21):6939–6944.

Yenish, J.P., D.A. Ball, and R. Schirman. 2009. Integrated management of jointed goatgrass in the Pacific Northwest. EB2042. Washington State University. Pullman, WA. Available online at http://cru.cahe.wsu.edu/CEPublications/EB2042/eb2042.pdf. Accessed November 18, 2009.

York, A.C., J.B. Beam, and A.S. Culpepper. 2005. Weed science: Control of volunteer glyphosate-resistant soybean in cotton. *Journal of Cotton Science* 9(2):102–109.

Zeimen, M.B., K.A. Janssen, D.W. Sweeney, G.M. Pierzynski, K.R. Mankin, D.L. Devlin, D.L. Regehr, M.R. Langemeier, and K.A. McVay. 2006. Combining management practices to reduce sediment, nutrients, and herbicides in runoff. *Journal of Soil and Water Conservation* 61(5):258–267.

Zelaya, I.A., and M.D.K. Owen. 2000. Differential response of common waterhemp (*Amaranthus rudis*) to glyphosate in Iowa. In *Proceedings of the Weed Science Society of America*, pp. 62–63 (Toronto, Canada). Weed Science Society of America.

Zelaya, I.A., and M.D.K. Owen. 2002. *Amaranthus tuberculatus* (Mq. ex DC) J.D. Sauer: Potential for selection of glyphosate resistance. In *Proceedings of the 13th Australian Weeds Conference*, Vol. 13, pp. 630–633 (Perth, Australia). Council of Australian Weed Science Societies.

Zelaya, I.A., M.D.K. Owen, and M.J. VanGessel. 2004. Inheritance of evolved glyphosate resistance in horseweed (*Conyza canadensis* (l.) Cronq.). *Theoretical Applied Genetics* 110:58–70.

———. 2007. Transfer of glyphosate resistance: Evidence of hybridization in *Conyza* (Asteraceae). *American Journal of Botany* 94(4):660–673.

Zhang, B.H., X.P. Pan, T.L. Guo, Q.L. Wang, and T.A. Anderson. 2005. Measuring gene flow in the cultivation of transgenic cotton (*Gossypium hirsutum* L.). *Molecular Biotechnology* 31(1):11–20.

Zhang, W., T.H. Ricketts, C. Kremen, K. Carney, and S.M. Swinton. 2007. Ecosystem services and dis-services to agriculture. *Ecological Economics* 64(2):253–260.

Zhao, J.Z., J. Cao, H.L. Collins, S.L. Bates, R.T. Roush, E.D. Earle, and A.M. Shelton. 2005. Concurrent use of transgenic plants expressing a single and two *Bacillus thuringiensis* genes speeds insect adaptation to pyramided plants. *Proceedings of the National Academy of Sciences of the United States of America* 102(24):8426–8430.

Zwahlen, C., W. Nentwig, F. Bigler, and A. Hilbeck. 2000. Tritrophic interactions of transgenic *Bacillus thuringiensis* corn, *Anaphothrips obscurus* (Thysanoptera: Thripidae), and the predator *Orius majusculus* (Heteroptera: Anthocoridae). *Environmental Entomology* 29(4):846–850.

Zwahlen, C., A. Hilbeck, R. Howald, and W. Nentwig. 2003. Effects of transgenic Bt corn litter on the earthworm *Lumbricus terrestris*. *Molecular Ecology* 12(4):1077–1086.

Zwahlen, C., A. Hilbeck, and W. Nentwig. 2007. Field decomposition of transgenic Bt maize residue and the impact on non-target soil invertebrates. *Plant and Soil* 300(1–2):245–257.

3

Farm-Level Economic Impacts

As shown in Chapter 1, farmers growing soybean, cotton, and corn adopted genetically engineered (GE) varieties over the last decade on the majority of acres planted to these crops in the United States. Much smaller acreages were planted in 2009 to a few other GE crops, such as canola, sugar beet, squash, and papaya. The decision to plant GE crops has affected the economic circumstances not only of the adopting farmers but in some cases of farmers who chose not to adopt them. The economic effects on farmers who adopt GE crops span their production systems and marketing decisions. In this chapter, we discuss the potential yield effects, changes in overhead expenses and management requirements, and shifts in market access and value of sales. A wide array of studies conducted mostly during the first 5 years of adoption has provided evidence for assessing the overall economic implications for farmers (see Box 3-1). We also discuss here the economic effects of GE-crop use on livestock producers who use the crops for feed and on farmers who do not elect to use the technology. The chapter concludes by examining the economic implications of gene flow from GE crops to non-GE crops and weedy relatives.

ECONOMIC IMPACTS ON ADOPTERS OF GENETICALLY ENGINEERED CROPS

GE crops have affected the economic status of adopters in several ways. The use of GE crops has had an effect on yields and their risk-management decisions. Genetic-engineering technology has also changed

BOX 3-1
Measuring Impacts

To evaluate the economic impacts of GE crops on adopters and non-adopters, the committee relied on the results of empirical analyses of farmer surveys and market data. Studies were peer reviewed, but the research approach and methods varied considerably with each study's purposes and data. Each study has its own strengths and limitations. For example, some studies may use a different guideline in judging the significance (i.e., confidence level) of factors affecting the adoption of GE crops compared to other studies. The committee could not make the various studies comparable and accepted each set of findings as valid evidence. Some of the general approaches used to estimate economic impacts are explained here.

Empirical data. A comparison of means or averages is sometimes used to analyze results from experiments in which factors other than the item of interest are "controlled" by making them as similar as possible. For example, means of yield or pesticide use can be compared for two groups of soybean plots that are similar in soil type, rainfall, sunlight, and all other respects. One of the two groups is considered to have a *treatment* (e.g., soybean with a genetically engineered trait), and the other does not (e.g., conventional soybean). As an alternative to controlled experiments, the subjects that receive treatment and those that do not can be selected randomly with data collected through mail, phone, Internet, or personal surveys.

Survey data. Caution must be exercised in interpreting the results obtained by analyzing the differences in means from data from "uncontrolled experiments," such as farm surveys. Conditions other than the "treatment" are not equal across the farms surveyed. For example, differences between mean estimates for yield and pesticide use from survey results cannot necessarily be attributed to the use of GE seeds because the different results are influenced by many other factors which are not controlled, including irrigation, weather, soil, nutrient and pest-management practices, other cropping practices, operator characteristics, and pest pressures.

Moreover, farmers are not assigned randomly to the two groups (adopters and nonadopters) but make the adoption choices themselves. Therefore, adopters and nonadopters may be systematically different as groups, and these differences may manifest themselves in farm performance. They could be confounded with differences due to the adoption of GE crops (i.e., the treatment). This situation, called self-selection, would bias the statistical results unless it is recognized and corrected.

However, farmer surveys give a more accurate picture of the total farm-level economic effects of GE-crop adoption in terms of the secondary behavioral changes resulting from adoption (e.g., adoption of conservation tillage and changes in the timing of pesticide application). Moreover, it is rarely the case that a farmer would or could choose to adopt a GE cultivar to replace a non-GE cultivar that is an isoline or near-isoline, so relying on agronomic experimental data to measure the economic differences can be biased. Also, only farmer surveys can reveal the value of the changes in nonpecuniary characteristics that can occur with the adoption of GE cultivars.

Social scientists often are able to statistically control for certain influencing factors for which there are data (apart from the GE-crop treatment) by using multiple regression techniques in econometric models. That is, differences in economic conditions and crop or management practices that also influence yield or other outcomes are held constant so that the effect of adoption can be isolated. For example, in research on GE crops, economists control for many factors, including output and input prices, pest infestation levels, farm size, operator characteristics, and management practices such as crop rotation and tillage. In addition, economists control for self-selection and simultaneity (of GE adoption and pesticide use decisions) using particular types of econometric models. To account for simultaneity of decisions and self-selectivity, a two-stage model may be used. The first stage consists of the adoption-decision model for GE crops. The second stage then uses the findings from the first stage to examine the impact of using GE crops on yield, farm profit, and pesticide use.

The Counterfactual. Ideally, measuring the impact of a treatment requires the observation of the results that would emerge in the absence of the treatment: a *counterfactual*. Aside from controlled experiments, it is not possible to observe this counterfactual outcome. Rather, the counterfactual is inferred by methods such as those summarized above (e.g., controlling for all other influencing factors). Moreover, regarding environmental impacts, Ferraro (2009) argues that "elucidating casual relationships through counterfactual thinking and experimental or quasi-experimental designs is absolutely critical in environmental policy and that many opportunities for doing so exist." The use of the two-stage estimation procedure to correct for selection bias exemplifies such a quasi-experimental design. However, Ferraro also admits that "not all environmental programs are amenable to experimental or quasi-experimental design." In those cases, firm conclusions cannot be drawn about the causative factors inducing GE-crop adoption or other outcomes of interest.

farmers' production expenses and altered their decisions related to time management. Furthermore, because of the widespread adoption of GE crops and their subsequent impact on yields, genetic-engineering technology has influenced the prices received by U.S. farmers.

Yield Effects

The first generation of GE varieties contains traits that control or facilitate the control of pest damage. A starting point for analyzing the productivity effect of such control is the damage-control framework (Lichtenberg and Zilberman, 1986) that was developed to estimate the effectiveness of the use of chemical pesticides and other pest-control activities. The framework recognizes that damage-control agents, like pesticides and GE traits for pest management, have an indirect effect on yield by reducing or facilitating the reduction of crop losses, in contrast with such inputs as fertilizers, capital, and labor, which affect yields directly. In particular, the framework assumes that

$$\text{effective yield} = (\text{potential yield})(1 - \text{damage}).$$

Potential yield is defined as the yield that would be realized in the absence of damage caused by pests (i.e., weeds, insects).[1] It is a function of production inputs, such as water and fertilizer, and of agroecological conditions and seed varieties. The yield actually observed is called effective yield and is equal to potential yield minus damage. Damage is affected by the pervasiveness of pests, which may be controlled with pesticides, the adoption of GE varieties, or other control activities. With that framework, the yield effects of GE varieties can be analyzed, but spatial, temporal, and varietal factors must be taken into consideration.

Indirect Yield Effects

The indirect yield effects of the use of insect-resistant (IR) crops are most pronounced in locations and years in which insect-pest pressures are high. For example, it is generally recognized that the adoption of Bt corn for European corn borer (*Ostrinia nubilalis*) control resulted in annual average yield gains across the United States of 5–10 percent (Falck-Zepeda et al., 2000b; Carpenter et al., 2002; Fernandez-Cornejo and McBride, 2002; Naseem and Pray, 2004; Fernandez-Cornejo and Li, 2005). Empirical

---

[1]Damage may also be caused by weather conditions, such as wind, rain, drought, and frost. For succinctness and convenience here, the definition of damage is restricted to pest problems.

studies, however, have clearly indicated that the indirect yield effects of Bt corn hybrids for European corn borer control vary temporally and spatially. In years with high pressure—corn borer damage of more than one tunnel per plant that exceeds 2 inches in length (Baute et al., 2002; Dillehay et al., 2004)—the yield advantage for Bt hybrids relative to near-isolines[2] was 5.5 percent in Pennsylvania and Maryland (Dillehay et al., 2004), 6.6 percent in Wisconsin (Stanger and Lauer, 2006), 8 percent in New Jersey and Delaware (Singer et al., 2003), 9.4 percent in Iowa (Traore et al., 2000), and 9.5 percent in South Dakota (Catangui and Berg, 2002). The yield advantage for Bt corn was negligible in those regions during years with low pest pressure (Traore et al., 2000; Catangui and Berg, 2002; Singer et al., 2003; Dillehay et al., 2004; Stanger and Lauer, 2006). Likewise, in regions where European corn borer is an occasional pest, there was no indirect yield advantage from the use of Bt hybrids in comparison to near-isolines (Cox and Cherney, 2001; Baute et al., 2002; Ma and Subedi, 2005; Cox et al., 2009). Most of the early empirical studies, however, included some Bt events[3] that did not have season-long control of corn borer, and this may have muted the yield advantage of Bt hybrids (Traore et al., 2000; Catangui and Berg, 2002; Pilcher and Rice, 2003).

There have been fewer empirical studies of the yield effects of Bt corn for control of corn rootworm (*Diabrotica* spp.) than of the effects of Bt corn for control of European corn borer. Rice (2004) estimated potential annual benefits if 10 million acres of Bt corn for corn rootworm control were planted. They included

- Intangible benefits to farmers (safety because of reduced exposure to insecticides, ease and use of handling, and better pest control).
- Tangible economic benefits to farmers ($231 million from yield gains).
- Improved harvesting efficiency due to reduced stalk lodging.
- Increased yield protection (9 to 28 percent relative to that in the absence of insecticide use and 1.5 to 4.5 percent relative to that with insecticide use).
- Reduction in insecticide use (a decrease of about 5.5 million pounds of active ingredient per 10 million acres).

---

[2]Near-isolines are cultivars that have the same or near genetic constitution (except for alleles at one or a few loci) as the original cultivar from which they were developed. Near-transgenic isolines that have similar genetic makeup except for the transgenic trait allow a comparison of the cultivar with or without the transgene for its agronomic, quality, or nutritional aspects.

[3]Each seed company has different events associated with different insertion places of the Bt gene and different promoter genes that allow a Bt toxin to be produced at different times of the season or in different plant parts.

- Increased resource conservation (about 5.5 million gallons of water not used in insecticide application).
- Conservation of aviation fuel (about 70,000 gallons not used in insecticide application).
- Reduced farm waste (about 1 million fewer insecticide containers used).
- Increased planting efficiency.
- Improved safety of wildlife and other nontarget organisms.

A recent study by Ma et al. (2009) indicated spatial and temporal variability in indirect yield responses. Bt corn rootworm hybrids produced yields 11–66 percent larger than untreated near-isoline hybrids. Bt yields were also larger than yields of the non-Bt hybrid variety planted on clay soils and treated with insecticide in 1 of 3 years that had high infestations of western corn rootworm (*Diabrotica virgifera virgifera*). On sandy soils, where corn rootworm infestations are typically much lower than on clay soils, yield differences also occurred between Bt corn rootworm hybrids and their near-isolines with or without the standard soil-applied insecticide treatment in 1 of 2 years. The study reported low levels of western corn rootworm on droughty sandy soil, however, and attributed yield increase to improved drought tolerance from the finer, longer fibrous roots of the Bt hybrid corn. Cox et al. (2009) found no yield advantage for corn hybrids with Bt rootworm control compared with near-isolines in a dry year when rootworm damage did not occur.[4]

Gray et al. (2007) expressed concern that one of the Bt corn rootworm events was somewhat susceptible to injury by a variant of western corn rootworm in Illinois. Another Bt corn rootworm event, however, had superior control of western corn rootworm larvae in Iowa, Illinois, and Indiana (Harrington, 2006); this suggests that distinct Bt events from different seed companies may differ somewhat in corn rootworm control as they did initially in corn borer control. Cox et al. (2009) evaluated both Bt rootworm events on second-year corn in field-scale studies on four farms

---

[4]As discussed in Chapter 1, all Bt rootworm corn hybrids are treated with a low level of insecticide and fungicide (typically a neonicotinoid). The low level (0.25 mg of active ingredient per seed) targets secondary pests but does not affect corn rootworm. In fields planted continuously with corn, the low level used with a soil-applied insecticide resulted in lower corn yields compared to a high level (1.25 mg of active ingredient per seed) with a soil-applied insecticide (Cox et al., 2007c). That is indirect evidence that the high level of seed-applied insecticide increases control of corn rootworm, but the low level does not. In addition, the low and high levels of seed-applied insecticides had no positive effects on corn grain (Cox et al., 2007b) or corn silage yields (Cox et al., 2007a) when following soybean, which suggests there is no yield enhancement of these seed-applied insecticides in the absence of pests.

in New York and found that neither rootworm event provided a yield advantage because rootworm occurrence was low in all fields. As with Bt corn for corn borer, Bt corn for rootworm control did not provide an indirect yield benefit in the absence of pest pressure.

Piggott and Marra (2007), relying on 1999–2005 university field-trial data from North Carolina, found that Bt cotton with two endotoxins out-yielded conventional cotton by 128 more lbs of lint per acre (14 percent of average yield in the region) and out-yielded Bt cotton with one endotoxin by 80 lb/acre (8 percent of average regional yield). A study of Bt cotton varieties with two endotoxins in 13 southern locations that had mostly moderate to high infestations of cotton bollworm (*Helicoverpa zea*), with or without foliar-applied insecticides, showed that indirect yield effects had spatial variability. The Bt cotton cultivars without insecticide use provided consistent control of the *Heliothines* (cotton bollworm and tobacco budworm, *Heliothis virescens*), regardless of the magnitude of infestation (Siebert et al., 2008). Furthermore, supplemental insecticide applications to the Bt cotton cultivars rarely improved control of budworm and bollworm. In the low-infestation environments, however, the use of Bt cultivars with or without insecticides provided no yield improvement relative to the control of the non-Bt cultivar without insecticide application. In the moderate- to high-infestation environments, the Bt cultivars provided the same 30-percent yield increase in lint yield with or without insecticides compared with the control (Siebert et al., 2008). In a large-scale study of 81 commercial cotton fields conducted in 2002 and 2003, average yield did not differ among Bt cotton, Bt cotton resistant to glyphosate, and non-GE cotton (Cattaneo et al., 2006). However, after statistical control for variation in two factors significantly associated with yield (number of applications of synthetic insecticide and seeding rate), the yield of Bt cotton and Bt cotton with herbicide resistance was significantly larger (by 8.6 percent) than the yield of non-GE cotton. A total of eight GE cotton cultivars and 14 non-GE cultivars were included in the study. For those cultivars, it appears that Bt cotton (herbicide-resistant or not) would generally out-yield non-Bt cotton given similar production inputs and agronomic conditions.

The indirect yield effects of herbicide-resistant (HR) crops generally may have less spatial and temporal variability because weeds are ubiquitous and cause yield losses in most situations. For example, the use of HR soybean with timely glyphosate application almost always achieves yield gains relative to production without weed control (Tharp and Kells, 1999; Corrigan and Harvey, 2000; Mulugeta and Boerboom, 2000; Wiesbrook et al., 2001; Knežević et al., 2003a, 2003b; Dalley et al., 2004; Scursoni et al., 2006; Bradley et al., 2007; Bradley and Sweets, 2008). Likewise, the use of HR corn and cotton varieties with timely glyphosate application almost always results in yield increases (Culpepper and York, 1999; Johnson et al., 2000; Gower et al., 2002; Dalley et al., 2004; Richardson et al., 2004; Sikkema et al., 2004; Thomas et al., 2004; Cox et al., 2005; Myers et al., 2005; Sikkema et al., 2005; Cox et al., 2006; Thomas et al., 2007).

Yield Lag and Yield Drag

Despite properties that result in indirect yield benefits, some farmers observed a yield reduction when they first adopted HR varieties (Raymer and Grey, 2003). Indeed, shortly after the adoption of glyphosate-resistant soybean, university soybean trials reported lower yields of HR varieties (Oplinger et al., 1998; Nielsen, 2000). In a study that compared five HR varieties with five non-HR varieties in four locations in Nebraska, evidence of "yield lag" and "yield drag" was found (Elmore et al., 2001a, 2001b).[5] A 5-percent yield lag was due to the difference in productivity potential between the older germplasm used to develop the HR varieties and the newer, higher yielding germplasm of the non-HR varieties.[6] A 5-percent yield drag resulted from the reduced production capacity of the soybean plant following the presence or insertion process of the HR gene (Elmore et al., 2001b). Although not as pronounced as in the Nebraska study, Bertram and Petersen (2004) also presented data that indicated a potential yield lag at one location in Wisconsin with the early HR soybean varieties.

Fernandez-Cornejo et al. (2002b) reported that a national farm-level survey indicated that HR soybean showed a small advantage in yield over conventional soybean, probably because of better weed control.

---

[5]Yield lag is a reduction in yield resulting from the development time of cultivars with novel traits (in this case, glyphosate resistance and Bt). Because of the delay between the beginning of the development of a cultivar with a novel trait and its commercialization, the germplasm that is used has lower yield potential than the newer germplasm used in cultivars and hybrids developed in the interim. Consequently, the cultivars with novel traits have a tendency to initially yield lower than new elite cultivars without the novel traits. Over time, the yield lag usually disappears.

Yield drag is a reduction in yield potential owing to the insertion or positional effect of a gene (along with cluster genes or promoters). This has been a common occurrence throughout the history of plant breeding when inserting different traits (e.g., quality, pest resistance, and quality characteristics). Frequently, the yield drag is eliminated over time as further cultivar development with the trait occurs.

[6]During selection for a particular trait in a plant-breeding program, many other traits may also change. Such "correlated" changes may occur because a gene controls more than one trait (pleiotropy), because genes controlling two traits are in physical proximity on a chromosome (linkage), or because of random segregation (drift). The distinctions among the three causes are important because the solutions to them differ. Solutions may be necessary because some correlated changes are undesirable.

A national survey of soybean producers in 2002 found that there was no statistical difference in yield between conventional soybean and HR soybean (Marra et al., 2004). A mail survey of Delaware farmers in 2001 found that HR soybean had a 3-bushel/acre yield advantage (Bernard et al., 2004). The survey data and results of empirical studies in Wisconsin indicate that the use of more elite germplasm in variety development has probably eliminated the yield lag or yield drag associated with the use of HR varieties (Lauer, 2006).

Similarly, early empirical studies of Bt corn hybrids indicated a potential yield lag, as indicated by the lower yield of Bt hybrids than of new elite hybrids (Lauer and Wedberg, 1999; Cox and Cherney, 2001). However, Bt hybrids yielded as well as or better than near-isolines (Lauer and Wedberg, 1999; Traore et al., 2000; Cox and Cherney, 2001; Baute et al., 2002; Dillehay et al., 2004), and this suggests that there was no yield drag or loss of yield because of the insertion of the Bt gene with the early Bt corn hybrids.

Furthermore, whether a yield loss or a yield increase materializes for a GE crop depends on the particular farming situation. For example, in their comparison of HR corn hybrids with non-HR varieties, Thelan and Penner (2007) reported that in low-yield environments HR hybrids yielded 5 percent more than non-HR hybrids and in high-yield environments non-HR hybrids yielded about 2 percent more than HR hybrids. An early study of cotton (May and Murdock, 2002) that compared first-generation glyphosate-resistant cultivars with nonresistant cultivars showed no yield lag in glyphosate-resistant cultivars and a yield advantage of using glyphosate instead of the standard conventional soil-applied herbicides. The results of the study suggested that the use of soil-applied herbicides resulted in some type of injury to cotton, whereas glyphosate application before the fourth leaf stage did not. A study at nine locations across the United States (May et al., 2004) showed that one of Monsanto's later glyphosate-resistant cotton lines provided even greater yield than the first-generation glyphosate-resistant cotton when glyphosate was applied from the fourth to the 14th leaf stage; this resulted in an agronomic advantage of the later technology.

A 2002 U.S. Department of Agriculture (USDA) survey found that increases in cotton yields in the Southeast were associated with the adoption of HR cotton and Bt cotton in 1997: A 10-percent increase in HR-cotton acreage led to a 1.7-percent increase in yield and a 10-percent increase in Bt cotton acreage led to a 2.1-percent increase in yield if other productivity-influencing factors were constant (Fernandez-Cornejo and Caswell, 2006).

It was noted above that most of the yield studies of GE versus non-GE crops conducted in the United States used data from the late 1990s and early 2000s.[7] Any yield differences between GE and non-GE varieties found during the first 5 years of adoption could have diminished as seed companies developed new IR and HR events. One reason for the lack of recent studies on yields may be that it is increasingly difficult to find sufficient data on non-GE varieties owing to the predominance of GE varieties in major crops (see Chapter 4).

Improved Crop Quality and Risk Management

Bt corn has been found to decrease concentration of the toxic chemical aflatoxin (Wiatrak et al., 2005; Williams et al., 2005) and some other mycotoxins produced by fungi (fumonisins in particular) in the grain (Clements et al., 2003). In doing so, it decreases the risk of price dockage to farmers because of poor crop quality and increases food safety for consumers. Bt crops also have reduced stalk lodging at harvest (Rice, 2004; Wu et al., 2005; Stanger and Lauer, 2006; Wu, 2006);[8] this improves crop quality and increases harvest efficiency, thus reducing the farmers' fuel and labor costs. A benefit of the use of HR soybean is that the presence of foreign matter (such as weed seeds) in the harvested crop has greatly decreased (from 5-25 percent to 1-2 percent in the southeastern states) (Shaw and Bray, 2003), reducing the need for handlers to blend soybean with high foreign matter with soybean with lower foreign matter to improve the overall quality of the crop.

The use of GE crops can also reduce agronomic risks for farmers. For example, in the case of HR crops, glyphosate breaks down quickly in the soil, removing the potential for the residual herbicide to injure a succeeding crop (Scursoni et al., 2006). Additionally, some Bt varieties may improve drought tolerance (Wilson et al., 2005). Empirical studies have not documented that the use of Bt corn for corn borer provides a yield benefit in the presence of drought (Traore et al., 2000; Dillehay et al., 2004; Ma et al., 2005), but Ma et al. (2009) found in an empirical study on Bt corn for corn rootworm that in a drought year on sandy soil, the Bt corn rootworm hybrid yielded 10 percent more than the near-isoline. The roots of the Bt corn rootworm hybrids were longer and more dense than those of the non-GE hybrid because the Bt trait kills the below-ground larvae that feed on the roots of the corn plant. Ma et al. (2009) speculated that Bt

---

[7]More recent data from field trials are available but have not been published in peer-reviewed literature.

[8]Stalk lodging is the permanent displacement of the stems of crops from their upright position, resulting in a crop that either leans or can be prostrate. A mildly lodged crop results in only a slight slowdown of harvest, whereas a severely lodged crop greatly slows down harvest (in some instances the crop can only be harvested in one direction, further reducing harvesting efficiency).

corn rootworm hybrids may have more drought tolerance than standard hybrids in drought years because the root system is more intact and therefore capable of taking up more water. Such risk reduction may explain in part farmers' motivation to adopt these GE crops. A related risk posed by adoption of Bt corn in northern latitudes, however, is the potential for higher grain moisture at harvest because of improved plant health, which increases drying costs or delays harvest (Pilcher and Rice, 2003; Dillehay et al., 2004; Ma and Subedi, 2005; Cox et al., 2009).

Because GE crops have the ability to reduce yield loss, adopting farmers also have different insurance options for managing risk. In 2007, Monsanto developed a submission to the USDA Federal Crop Insurance Corporation for a new crop-insurance endorsement for corn that contains three traits: a Bt toxin that controls corn borer, one that controls corn rootworm, and herbicide resistance.[9] The submission proposed a premium-rate discount for those hybrids based on several thousand on-farm field trials conducted over several years in the Corn Belt states of Illinois, Indiana, Iowa, and Minnesota. The trials demonstrated the yield and yield-risk reduction advantages of the hybrids compared with conventional or single-trait HR hybrids and showed that the current premium rates were no longer actuarially appropriate. A lower insurance premium became available in the 2008 crop year to farmers who adopted the triple-stacked hybrids. The rate discount was applied to the yield portion of the premium for actual production history of the field and based policies on crop-insurance units in which at least 75 percent of the acreage was planted to qualifying corn hybrids. The average premium-rate discount was 13 percent in 2008, or about $3.00/acre.

Comparable triple-stacked hybrids from seed companies Dupont/Pioneer and Syngenta were approved for inclusion in the program for the 2009 crop year, and the premium-rate discount applies to all three companies' and licensees' seed brands that contain at least the above-mentioned traits for dryland corn in at least a subset of 13 Midwest states and irrigated corn in Kansas and Nebraska. This is the first approved crop-insurance innovation that has resulted in reduced premium rates, and it provides a saving for farmers and reduces the need for premium subsidies by the federal government. Cox et al. (2009), however, found no consistent yield or economic advantage for triple-stacked hybrids compared to double-stacked hybrids from both companies in second-year corn in New York, despite one of the years being dry and warm. In both years, corn rootworm damage was low, and corn borer damage was sporadic across locations.

---

[9]These products are marketed by Monsanto as YieldGard® Plus, Roundup Ready 2®, and YieldGard VT Triple® hybrids.

Production Expenses

The use of GE crops triggers changes in several production expenses, particularly those related to seed technology, pesticide expenditures, labor and management requirements, and machinery operations.

Seed Prices

U.S. farmers pay for the GE traits in the seeds that they plant in the form of a technology fee because GE seeds are considered proprietary in the United States. The market price of seed, which includes the technology fee, incorporates the costs associated with development, production, marketing, and distribution (Fernandez-Cornejo and Gregory, 2004). The price must be responsive to farmers' willingness to purchase the technology while ensuring an attractive return on capital to the seed development firms (technology provider and licensee seed companies or distributors) and their investors. The price also depends on the competitiveness of the particular seed market and on the pricing behavior of firms that hold large shares of the market.

In recent decades, private-sector research and development costs have risen with the application of new technologies. Much of the increase in seed prices paid by U.S. farmers has been associated with that general trend (Krull et al., 1998). The seed-price index has exceeded the average index of prices paid by U.S. farmers by nearly 30 percent since the introduction of GE seeds in 1996 (Figure 3-1). The contrast is even starker for cotton and soybean. After adjustment for inflation, the real average cotton seed price almost tripled between 1996 and 2007 (Figure 3-2), while the soybean seed price grew by more than 60 percent.

The rise in real seed prices can be accounted for by improvements in germplasm, by the increasing price premiums paid for GE seed, and by the growing share of GE seed purchased by U.S. farmers (as the share of seed saved by farmers correspondingly decreased). The price premium, which includes the technology fee, doubled in real terms for GE cotton seed between 2001 and 2007 (adjusted for inflation) (Figure 3-3). U.S. farmers also experienced similar price-premium increases for corn and soybean seed (Figures 3-4 and 3-5). Some of the increase reflects the larger number of services that the seed delivers to the buyers compared with conventional seed. For example, farmers who purchase Bt cotton receive the seed germplasm and an insecticide combined in one product, whereas for non-GE crops they must buy each separately and pay for costs related to applying the insecticide. The increase also reflects the additional value to the farmer provided by later GE cultivars with more than one type of trait or more than one mode of action for particular target pests. The rates of adoption noted in Chapter 1 indicate that the price premiums have not

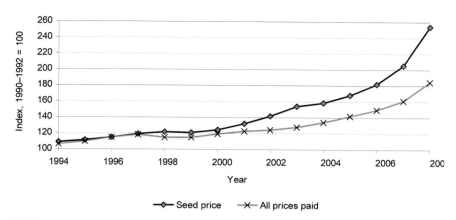

FIGURE 3-1 Seed-price index and overall index of prices paid by U.S. farmers.
SOURCE: Fernandez-Cornejo, 2004; USDA-NASS, 2000, 2005, 2009a.

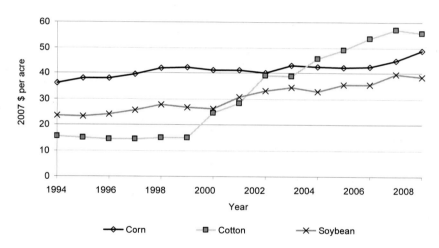

FIGURE 3-2 Estimated average seed costs for U.S. farmers in real (inflation-adjusted) terms.
SOURCE: Fernandez-Cornejo, 2004; USDA-NASS, 2000, 2005, 2009a.

deterred many U.S. farmers from purchasing GE seeds and that non-GE seed options were less attractive or were not available.

Other Input Costs

If U.S. farmers who have adopted GE crops pay higher prices for the seed, have they experienced compensatory cost reduction for other

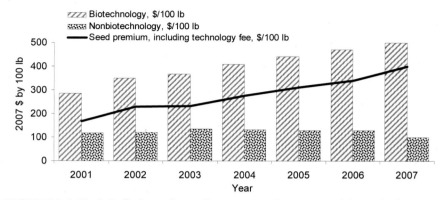

FIGURE 3-3 Real (inflation-adjusted) cotton seed prices paid by U.S. farmers, 2001–2007.
SOURCE: USDA-NASS, 2000, 2005, 2009a.

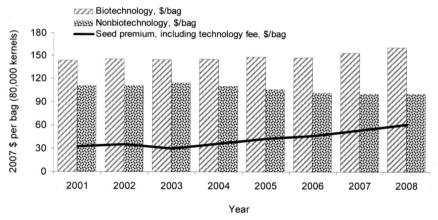

FIGURE 3-4 Real (inflation-adjusted) corn seed prices paid by U.S. farmers, 2001–2008.
SOURCE: USDA-NASS, 2000, 2005, 2009a.

inputs? With insect-resistance technology, a plant contains its own insecticide, whereas most HR crops are engineered to be used with the herbicide glyphosate. Have those conditions changed adopters' farming practices and purchasing habits?

Economic reasoning suggests that the influence of genetic engineering on pesticide use depends on whether the GE cultivar and the pesticides are complementary or substitute inputs (Just and Hueth, 1993). Where IR or stacked GE cultivars substitute for other pesticides, chemical-pesticide

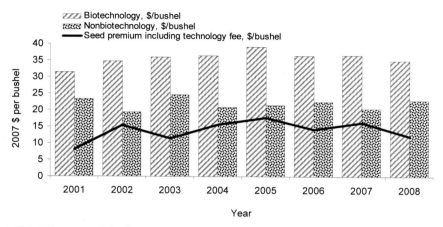

FIGURE 3-5 Real (inflation-adjusted) soybean seed price paid by U.S. farmers, 2001–2008.
SOURCE: USDA-NASS, 2000, 2005, 2009a.

use should decline. That is often the case with Bt crops (see Chapter 2, Figures 2-7 and 2-8). For HR crops, it often means reducing the use of less effective, more costly, and possibly more toxic herbicides although exceptions occur (Cattaneo et al., 2006). That substitution effect can produce cost savings as well as reductions in environmental and human health risks associated with chemical applications (Sydorovych and Marra, 2007). Several studies have attempted to establish whether the adoption of GE crops affects pesticide use. Some early investigations found evidence of a decline in pesticide use as adoption of GE crops increases (Heimlich et al., 2000; Hubbell et al., 2000; Carpenter et al., 2001; Marra et al., 2002). Some studies have found that most of the reduction in pesticide use resulting from adoption of an IR cultivar was in highly toxic chemicals, and average toxicity declined with adoption (Heimlich et al., 2000; Sydorovych and Marra, 2007). However, others have concluded that pesticide use increases in tandem with GE-crop production (Benbrook, 2004). Such contradictory findings have been attributed to the different approaches to measuring pesticide use, specifically

• How pesticide use is recorded (pesticide active-ingredient volume, formulated volume, relative toxicity, or number of applications) (Sydorovych and Marra, 2007).
• Which factors are controlled for (results would vary from region to region and from year to year depending on the extent of pest infestation, weather, cropping patterns, and so on).
• The method of aggregation (Frisvold and Marra, 2004).

A general overarching effect cannot be discerned because of the variability in specific conditions on different farms and in different regions.

The observed change in pesticide use with IR crops depends on the crop and the pest. Changes in insecticide use for treatment of European corn borer were minimal because many farmers accepted yield losses rather than incur the expense and uncertain results of chemical control. A survey of corn growers in Iowa and Minnesota determined that only 30 and 17 percent, respectively, had managed European corn borer with insecticides during any season in the early 1990s because chemical use was not always profitable and timely application was difficult owing to the unpredictability of pest outbreaks (Rice and Ostlie, 1997).

In the case of Bt cotton, however, GE control greatly reduced expenditures on pesticides to treat tobacco budworm, pink bollworm (*Pectinophora gossypiella*), and cotton bollworm (Jackson et al., 2003; Cattaneo et al., 2006). Survey data indicated that the number of insecticide sprays and insecticide costs generally decreased with the adoption of GE cotton across the United States (Table 3-1). Where measurable, farm-level profit was also shown to have increased with the adoption of Bt cotton in all states (Piggott and Marra, 2007). Although the studies reported in Table 3-1 seem to suggest that insecticide costs increased after commercialization of Bt cotton in Arizona, detailed surveys of insecticide use and costs conducted since 1979 clearly show that use and costs were drastically reduced after 1996 (Ellsworth et al., 2009). One major factor in the reductions has been the efficient control of pink bollworm by Bt cotton (Carrière et al., 2003, 2004). However, other critical factors in reducing insecticide use and cost were the introduction of novel and highly efficient insecticides for the control of whitefly (*Bemisia tabaci*) in cotton (Carrière et al., 2004) and the success of the boll weevil eradication program (Fernandez-Cornejo et al., 2009). That illustrates an important point (see Chapter 2): Longitudinal data on pesticide use should not be taken at face value in assessing the effects of GE crops without controlling for other influences, as many factors can contribute to changes in patterns of insecticide use.

Bt corn is a preferred method for growers for controlling rootworm because of its simplicity and safety in applying it compared with soil-applied insecticides or with higher levels of active ingredient in seed treatments on non-Bt corn seed[10] (Al-Deeb and Wilde, 2005; Vaughn et al., 2005; Ahmad et al., 2006). The adoption of Bt corn for rootworm control

---

[10]As discussed in Chapter 1, Bt corn hybrid seed for corn rootworm control has 0.25 mg of active ingredient of insecticide and fungicide applied per seed compared to 1.25 mg of active ingredient applied to non-Bt corn hybrids.

has resulted in a substantial reduction in insecticide use, by an estimated 5.5 million pounds of active ingredient per 10 million acres (Rice, 2004).

In addition to the pesticide quantity effects, the adoption of HR and IR crops lowers the demand of competing pesticides used on conventional varieties and may therefore lower the prices of these pesticides. Huso and Wilson (2006) shows that this effect benefits farmers who adopt the GE variety and those who plant the conventional variety.

Indirect cost differences between GE crops and conventional crops originate in the adoption of practices that are linked to the adoption of some GE crops. For example, if a GE crop reduces the need for till-age to control weeds, reductions in machinery, fuel, and labor for the avoided cultivation practices amount to indirect cost savings. The indirect cost differences are particularly important for HR crops because of the complementary relationship between their adoption and conservation tillage. That is, GE-crop adoption increased the probability of adoption of conservation tillage, and conservation tillage increased the probability of higher adoption of GE crops (for a more detailed discussion of conserva-tion tillage, see Chapter 2).

The increased use of conservation tillage has been facilitated by the commercialization of more effective postemergence herbicides, such as glyphosate, that can be applied topically to crops and weeds. Glyphosate can supplement or replace tillage as a tool for controlling most weeds and in so doing can reduce the use of machinery and fuel and lower labor requirements (Harman et al., 1985; Chase and Duffy, 1991; Baker et al., 1996; Downs and Hansen, 1998; Boyle, 2006; Baker et al., 2007). Mitchell et al. (2006) reported that a reduced-tillage system in the plant-ing of California cotton reduced the number of tractors in operation by 41 to 53 percent, fuel use by 48 to 62 percent, and overall production costs by 14 to 18 percent. Sanders (2000) reviewed and summarized results of several studies and concluded that conservation tillage can reduce fuel costs by as much as 50 percent and labor costs by up to 40 percent. Those conclusions agree with USDA Natural Resources Conservation Service estimates that Iowa farmers would save 30–50 percent in fuel costs by adopting conservation-tillage practices (Table 3-2). Using Nebraska sur-vey data for various row crops, Jasa (2000) showed that fuel use for no-till was 1.43 gal/acre compared with 5.28 gal/acre for moldboard-plow tillage and that labor requirements for no-till were 0.49 hours/acre, com-pared with 1.22 hours/acre for moldboard-plow use.

The financial returns to GE crops should vary directly with fuel prices if they save costly machinery passes over a field. HR crops do not necessarily save passes over a field, but they do substitute herbi-cide applications for more expensive and more fuel intensive methods of weed management, such as intensive tillage practices or the use of

TABLE 3-1 Summary of Farm-Level Impact Evidence for Genetically Engineered Cotton in the United States, 1996–1999

	Differences in:							
	Yield				Pesticide Cost			
Transgene Type, State	Number of estimates	Mean	Minimum	Maximum	Number of estimates	Mean	Minimum	Maximum
	(count)	(pounds lint per acre)			(count)	(dollar/acre)		
Bt cotton								
Alabama	4	143.5	231.5	38.0	2	−32.4	3.1	−68.0
Arizona	8	116.7	917.0	−331.5	9	17.1	97.0	−24.6
Georgia	3	75.2	104.0	38.0	3	−23.4	27.5	−68.0
Louisiana	2	−7.5	22.0	−37.0	2	−20.0	−15.4	−24.6
Mississippi	8	22.6	92.0	−73.0	8	−5.1	13.8	−24.6
North Carolina	8	41.6	182.5	−35.7	2	−14.3	−1.2	−27.5
Oklahoma	4	168.0	203.0	123.0				
South Carolina	2	90.5	119.0	62.0	2	−16.2	−1.2	−31.1
Tennessee	2	−79.0	85.0	−243.0	1	−5.6		
Texas	3	116.6	177.5	81.0				
Virginia	1	62.0			1	−1.2		
RR[a] cotton								
Arkansas	1	−150						
Tennessee	1	−243			1	−145.3		
Bt/RR cotton								
Arkansas	2	292.8	−331.5	917.0	2	79.5	−269.0	159.0

[a]Monsanto markets its glyphosate-resistant varieties under the trademarked name Roundup Ready.
SOURCE: Adapted from Marra, 2001.

herbicides that require physical incorporation into the soil. Also, with potentially fewer passes over the field, tractor and spraying equipment lasts longer, and this results in savings in machinery and equipment costs over the long term.

Management Requirements and Nonpecuniary Benefits

Many of the commercially available GE products have consistently been shown to be profitable for U.S. farmers. For example, the profitability of Bt cotton in the Cotton Belt and Bt corn for controlling corn rootworm is well documented (Marra, 2001; Alston et al., 2002). However, the national evidence supporting the use of HR soybean is inconclusive (Bullock and

Pesticide Use				Profit			
Number of estimates	Mean	Minimum	Maximum	Number of estimates	Mean	Minimum	Maximum
(count)	(sprays/acre)			(count)	(dollars/acre)		
2	−1.3	0.3	−3.0	2	77.6	116.5	38.7
3	−2.2	−1.8	−2.5	10	57.5	465.0	−104.0
3	−2.7	−2.5	−3.0	3	92.0	169.2	38.7
2	−2.4	−2.2	−2.5	2	16.5	36.0	−3.1
4	−2.4	−1.3	−3.3	6	34.5	79.5	−3.1
2	−2.4	−2.4	−2.5	8	20.5	95.1	−25.3
4	−3.4	−2.3	−6.5	4	53.8	85.5	25.5
2	−2.5	−2.5	−2.5	4	51.8	80.1	17.1
1	−1.8			2	67.5	74.3	60.7
				1	46.0		
1	−2.5			1	41.7		
				1	17.1		
				1	74.3		
				2	243.0	21.0	465.0

TABLE 3-2 Fuel Consumption by Tillage System (Gallons per Year)

Crop	Acres	Conventional Tillage	Mulch Till	Ridge Till	No-Till
Corn	1,000	4,980	3,710	3,330	2,770
Soybean	1,000	4,980	3,110	3,330	1,970
Total fuel use		9,960	6,820	6,660	4,740
Potential fuel savings over conventional tillage			3,140	3,300	5,220
Saving			32%	33%	52%

SOURCE: USDA-NRCS, 2008.

Nitsi, 2001; Gardner and Nelson, 2007). Fernandez-Cornejo et al. (2002b) and Fernandez-Cornejo and McBride (2002) evaluated 1997 field-survey data and 1998 whole-farm survey data, respectively, and found that the differences in net returns between adopters and nonadopters of HR soybean were not significant. This lack of significance is consistent with findings from other producer surveys (Couvillion et al., 2000; Duffy, 2001). In light of high overall adoption rates, those findings suggest that other considerations have motivated farmers to use genetic-engineering technology. The wide adoption of HR soybean despite the associated technology fee stimulated research to identify possible nonpecuniary benefits to GE adopters that motivate such a shift in technology use.

In addition to the substantially superior control of a broad spectrum of weeds (Scursoni et al., 2006), simplicity, flexibility, and increased worker safety have been suggested as root causes of herbicide-resistance technology adoption, in that growers can use one herbicide instead of several to control a wide array of broadleaf and grass weeds (Gianessi and Carpenter, 1999; Bullock and Nitsi, 2001). The convenience of HR soybean use may mean that farmers can reduce the time that they spend scouting fields for weeds and mixing and spraying different herbicides to address various weed problems (Bullock and Nitsi, 2001). Furthermore, the window of application for glyphosate is wider than that for other postemergence herbicides. That application flexibility can effectively control weeds but often the weeds have already caused a loss in potential crop yield by the time glyphosate is applied.

However, quantifying the simplicity, flexibility, and safety of pest-control programs has been difficult. The inability to include a measure of management time in the evaluation of benefits of new technologies in agriculture is not unique to HR soybean. As Fernandez-Cornejo and Mishra (2007) observed, assessments of technology adoption using traditional economic tools pioneered by Griliches (1957) have proved insufficient to explain differing rates of adoption of many recent agricultural innovations. The standard measures of farm profits, such as net returns to management, give an incomplete picture of economic returns because they usually exclude the value of management time itself (Smith, 2002). HR soybean was adopted rapidly despite showing no statistically significant advantage in net returns over conventional soybean in most studies, but adoption of such strategies as integrated pest management has been rather slow even though it has explicit economic and environmental advantages (Fernandez-Cornejo and McBride, 2002; Smith, 2002). That inconsistency led to the hypothesis that HR adoption is driven by unquantifiable advantages—such as presumed simplicity, flexibility, and safety—that translate into a reduction in managerial intensity, which frees time for other pursuits, and into increased worker safety.

An obvious use of managers' time is off-farm employment; alternatively, a farmer could farm more acreage to increase farm income. Fernandez-Cornejo and collaborators examined the interaction of off-farm income-earning activities and adoption of different agricultural technologies of varied managerial intensity, including HR crops (Fernandez-Cornejo and Hendricks, 2003; Fernandez-Cornejo et al., 2005) and Bt corn (Fernandez-Cornejo and Gregory, 2004). They also estimated empirically the relationship between the adoption of those innovations and farm household income from on-farm and off-farm sources. To do that, they expanded the agricultural household model to include the technology-adoption decision and off-farm work-participation decisions by the operator and spouse (Fernandez-Cornejo et al., 2005; Fernandez-Cornejo and Mishra, 2007).

Those studies hypothesized that adoption of management-saving technologies frees operators' time for use elsewhere, most notably in off-farm employment, and that leads to higher off-farm income. They found that the relationship between the adoption of HR soybean and off-farm household income is positive and statistically significant: After controlling for other factors, a 15.9-percent increase in off-farm household income is associated with a 10-percent increase in the probability of adopting HR soybean. The adoption of HR soybean is also positively and significantly associated with total household income from off-farm and on-farm sources. A 9.7-percent increase in total household income is associated with a 10-percent increase in the probability of adopting HR soybean. In contrast, and consistent with the lack of higher returns from this technology, adoption of HR soybean did not have a significant relationship with household income from farming. Those findings complement the findings of Gardner and Nelson (2007), who used national survey data from 2001–2003 and found that adopting HR soybean reduced household labor requirements by 23 percent.

Studies have also found that farmers value the convenience and reduced labor requirements of Bt cotton above and beyond the pecuniary benefits. Because conventional cotton faces heavy pest pressure, IR varieties decrease the time demands of spraying, and this leads to a 29-percent reduction in household labor requirements (Gardner and Nelson, 2007). Survey data of Marra and Piggott (2006) support the finding that farmers place a monetary value on the convenience, flexibility, and relative safety of GE crops. In a stated-preference approach, participants in four surveys placed values on such characteristics as saved time, operator and worker safety, and total convenience. In each survey that evaluated the total-convenience attribute of genetic-engineering technology, it made up over 50 percent of the total value placed on nontraded aspects of the GE crop (Table 3-3). The median total value of convenience ranged from $3.33/acre

TABLE 3-3 Value and Relative Importance of Nonpecuniary
Benefits to Farmers

| Characteristic | Rescaled[a] | | | |
	Median (%)	Mean (%)	Std Dev (%)	Share (%)
Corn Rootworm Survey: n = 367				
Time saving	0.588	0.997	1.390	23.86
Equipment saving	0.400	0.724	0.969	17.51
Operator and worker safety	0.429	0.991	1.623	17.12
Environmental safety	0.208	0.787	1.565	10.88
More consistent stand	0.800	1.773	2.862	30.63
Sum of the parts	**3.000**	**5.272**	**6.222**	
National Soybean Survey: n = 113				
Operator and worker safety	0.913	1.660	2.026	20.97
Environmental safety	1.304	1.961	2.201	24.89
Total convenience	3.333	4.158	3.690	54.14
Sum of parts	**5.000**	**7.779**	**6.026**	
North Carolina Herbicide-Resistant Crop Survey: n = 52				
Operator and worker safety	2.361	2.923	2.783	23.91
Environmental safety	1.666	2.720	2.660	20.45
Total convenience	5.000	7.793	7.818	55.63
Sum of parts	**10.000**	**13.437**	**10.612**	
Roundup Ready® Flex Cotton Survey: n = 72				
Operator and worker safety	1.875	3.056	4.061	23.90
Environmental safety	0.958	2.592	3.382	18.06
Total convenience	5.000	11.180	15.441	58.04
Sum of parts	**10.000**	**16.828**	**17.383**	

[a]Rescaled to conform the magnitude of the overall value, which is asked as a separate question.
SOURCE: Marra and Piggott, 2006.

per year for soybean to $5.00/acre per year for HR cotton. Survey respondents also placed a value on the improved operator and worker safety characteristics of GE crops. Farmers valued the reduction in handling and toxicity of the pesticides involved with those crops at $0.43–$2.36/acre per year. Although initially they increase the demand for GE seeds, the perceived management benefits may cause the demand for the seeds to become more inelastic (i.e., less responsive to price increases) over time. As farmers get accustomed to the characteristics and continue to place a value on them, increases in seed expenditures through either a price increase or an increase in user cost will not reduce the use of GE seeds by as large an amount as when the nonpecuniary attributes are not present (Piggott and Marra, 2008).

Management benefits do not appear to influence the adoption of GE corn. Fernandez-Cornejo and Gregory (2004) did not find a statistically significant relationship between adoption of Bt corn (to control corn borer) and off-farm household income, and Gardner and Nelson (2007) noted no effect of adoption of Bt or HR corn on household labor. The lack of a significant relationship supports the observation that most farmers accepted yield losses rather than incur the expense and uncertainty of chemical control for European corn borer before the introduction of Bt corn (Fernandez-Cornejo et al., 2002a). For those farmers, the use of Bt corn was reported to result in yield gains rather than pesticide-related savings, and savings in managerial time were small. However, one of the benefits of adoption of Bt corn for rootworm control is that it makes it unnecessary to handle toxic insecticides at planting or to deal with high rates of insecticide-treated seeds.

Thus, the econometric results are consistent with anecdotal statements that many GE crops save managerial time because of the associated simplicity and flexibility of pest control. In the case of some GE crops, such as HR soybean, these nonpecuniary benefits provide incentives for adoption that counteract the additional cost of GE seeds. Indeed, the benefits increase demand for the GE seeds, and that in turn supports a higher price. In the case of other GE crops, such as Bt cotton, nonpecuniary benefits are accrued above and beyond additional farm profits.

Lower management costs and increased yield and nonpecuniary benefits have figured in the economic value of the natural refuge for cotton with two endotoxins for control of the bollworm-budworm complex. As discussed in Chapter 2, the Environmental Protection Agency (EPA) changed the refuge requirement for these IR cotton varieties from a 20-percent refuge treated with insecticide (or a 5-percent refuge not treated with insecticides) to a natural refuge where wild host plants constitute the refuges. The benefits of the refuge change were estimated for North Carolina to be $26.90 per year per impacted acre when pecuniary and nonpecuniary impacts were considered (Piggott and Marra, 2007).

Value and Market-Access Effects

In addition to the input-cost effects, the use of GE crops can affect the revenue potential of farmers. Two such effects can occur: foreign yield effects on the prices of products sold and market access to sell the GE crops.

Increase in grain and oilseed supplies should result in downward pressure on the prices received by farmers, all else being equal. Genetic-engineering technology helped to boost yields that had already been growing over the last 70 years through improved plant-breeding techniques.

As a result, supply exceeded demand; the real price of food (adjusted for inflation) had fallen until 2006. However, over the period 2006–2008, corn and soybean prices increased rapidly because of various factors, including the rise in world incomes and the demand for renewable fuels made from agricultural feedstock. The increase in the global supply of those crops due to the adoption of GE crops and improvements in germplasm and plant breeding likely moderated the upward pressures on prices during this time.

Assessing the impact of new agricultural technologies on commodity prices is difficult because the effect on price cannot be measured directly. As Price et al. (2003) explain, once a new technology is introduced and adopted, only the world price that results from increasing global supply (supply shift) can be observed. It is not possible to observe the counter-factual price—the price that would have existed, assuming the same supply and demand conditions, if the new technology had not been introduced (see Box 3-1). Therefore, the counterfactual world prices and demanded quantities of the commodities must be estimated from market equilibrium conditions by using econometric models, which generally are reliable in the short term and when systems are stable.

The approach to calculating the effect of genetic-engineering technol-ogy on commodity prices followed by most studies (Falck-Zepeda et al., 2000a, 2000b; Moschini et al., 2000; Price et al., 2003; Qaim and Traxler, 2005) is based on the theoretical framework developed by Moschini and Lapan (1997) to assess the impacts of an innovation on economic welfare when the innovator behaves as a monopolist under the protection of intel-lectual-property rights in an input market by pricing the new technology above marginal cost (the cost of producing one more unit of a good) (Price et al., 2003). Changes in the economic welfare of producers and consumers in a competitive output market can also be measured because some of the benefits generated by the innovation are passed on to them in the form of higher production efficiency and lower commodity prices.[11]

---

[11]As Price et al. (2003) described it, the estimated total market benefit of adopting each of the GE crops depends on the extent to which the global commodity supply curve shifts outward after the introduction of the technology. In each case, the shift in supply reflects potential yield increases and/or decreases in costs. The estimated market benefit also depends on the interaction of the supply and demand curves before and after the introduction of the new technology. The empirical models calculate the preinnovation and postinnovation prices and quantities in an international market setting by using information on adoption rates, crop yields, costs, technology fees, and seed premiums. The framework also takes into account the adoption of biotechnology outside the United States. The counterfactual world price, the equilibrium world price without the innovation, is the sum of the observed market price and the vertical supply shift resulting from the adoption of GE crops. The equilibrium world price occurs at the intersection of the excess-supply and excess-demand curves.

Table 3-4 shows the estimates of the effect of GE-crop adoption (corn, soybean, and cotton) on crop prices. The price effects are different for each crop and technology and depend on the market penetration (the extent of adoption) of the new technology and on the details of the models used (particularly supply and demand parameters). For example, adoption of Bt cotton was associated with a decline in cotton prices of 0.65 percent for the first year of adoption in the United States only but with a price decline of 1.1 percent when adoption continued in the United States and took place in other countries. The effect of adoption of HR soybean varieties on soybean price ranged from a decline of 0.17 percent in 1997, when adoption had only occurred in the United States, to about a 2-percent decline following further adoption in the United States and Argentina and a 2.6-percent decline for world adoption in 2001. Simultaneous adoption of Bt corn and HR soybean could lead to a decline in corn prices of 2.5 percent and a decline in oilseed prices of 3.9 percent, all other things being equal.

Table 3-5 presents the estimated distribution of the tangible benefits among consumers, farmers, technology providers (biotechnology firms), seed firms, and consumers and producers in the United States and the rest of the world. The distribution of benefits varies by crop and technology because the economic incentives to farmers (crop prices and production costs), the payments to technology providers and seed firms, and the effect of the technology on world crop prices are different for each crop and technology. For example, farmer adoption of HR cotton benefits mainly consumers, whereas adoption of Bt cotton benefits farmers and technology providers. Innovators (technology providers and seed firms) are often the largest beneficiaries in the case of HR soybean, but producers and consumers also gain (Moschini et al., 2000). The aggregate net benefits to crop farmers depend on the aggregate cost saving relative to the estimated price decreases and increased production (sales). The lower output prices may deter some farmers who have relatively lower yield gains or higher costs from adoption. But farmers with sufficient yield gains and cost saving will adopt GE crops even when an increase in supply puts downward pressure on prices. Livestock producers primarily receive benefits from lower prices of feedstocks than would have occurred without GE-crop adoption. Analyses of the benefits of GE crops and their distribution have many nuances.

The studies mentioned above analyzed the economic effects of GE varieties during the early period of their adoption of these technologies (the latest study used data from 2001).[12] Results of studies of adoption in

---

[12]These studies were carried out before Brazil began producing large amounts of GE soybean. The entry of Brazil into the GE soybean market and the continued expansion of GE soybean in Argentina may have pushed considerable amounts of the benefits from producers to consumers.

TABLE 3-4 Effect of Global Adoption of Genetically Engineered Crops on Commodity Prices

Technology and Crop	Adopting Areas	Commodity Name	Price Decline (%)	References
Bt cotton	United States (first year)	Cotton	0.65	Falck-Zepeda et al., 2000b
Bt cotton	United States, other producing countries[a]	Cotton	1.11	Price et al., 2003
HR cotton	United States	Cotton	3.4	Price et al., 2003
HR soybean	United States	Soybean	0.17	Price et al., 2003
HR soybean	United States	Soybean	1.0	Moschini et al., 2000
HR soybean	United States, South America	Soybean	2.2	Moschini et al., 2000
HR soybean	World	Soybean	2.6	Moschini et al., 2000
HR soybean	United States, Argentina	Soybean	1.96[c]	Qaim and Traxler, 2005
HR soybean and canola, Bt corn[b]	United States, Canada, Argentina	Oilseeds	2.9	Anderson and Jackson, 2005
HR soybean and canola, Bt corn[b]	United States, Canada, Argentina	Corn	1.94	Anderson and Jackson, 2005
HR soybean and canola, Bt corn[b]	World	Oilseeds	3.08	Anderson and Jackson, 2005
HR soybean and canola, Bt corn[b]	World	Corn	2.09	Anderson and Jackson, 2005
Bt corn, HR soybean	United States Canada	Oilseeds	1.5	Fernandez-Cornejo et al., 2007
Bt corn, HR soybean	United States Canada	Corn	2.5	Fernandez-Cornejo et al., 2007
Bt corn, HR soybean	World	Oilseeds	3.87	Fernandez-Cornejo et al., 2007

[a]Assumes that countries other than the United States would have a 50-percent efficiency of technology transfer. Adoption is for 1997.
[b]Adoption rates assumptions vary by country and crop. See details in Anderson and Jackson (2005).
[c]Price decline reported is for 2001. Qaim and Traxler (2005) also calculated price declines for other years; for example, 1.54 percent for 2000 and 1.25 percent for 1999.

TABLE 3-5 Adoption of Genetically Engineered Crops and Their Distribution

Study	Year	Total Benefits ($ million)	Share of Total Benefits (%)			
			U.S. Farmers	Innovators	U.S. Consumers	Net RoW
Bt cotton						
Falck-Zepeda et al., 1999	1996	134	43	47	6	
Falck-Zepeda et al., 2000a	1996	240	59	26	9	6
Falck-Zepeda et al., 2000b	1997	190	43	44	7	6
Falck-Zepeda et al., 1999	1998	213	46	43	7	4
Frisvold et al., 2000	1996–1998	131–164	5–6	46	33	18
US-EPA, 2001[a]	1996–1999	16–46	NA	NA	NA	NA
Price et al., 2003	1997	210	29	35	14	22
Herbicide-resistant cotton						
Price et al., 2003	1997	232	4	6	57	33
Herbicide-resistant soybean						
Falck-Zepeda et al., 2000b	1997-LE[b]	1,100	77	10	4	9
	1997-HE[c]	437	29	18	17	28
Moschini et al., 2000	1999	804	20	45	10	26
Price et al., 2003	1997	310	20	68	5	6
Qaim and Traxler, 2005	1997	206	16[d]	49	35	NA[e]
Qaim and Traxler, 2005	2001	1,230	13[d]	34	53	NA[e]

NOTE: NA = not applicable; RoW = rest of the world (includes consumers and producers).

[a]Limited to U.S. farmers.

[b]LE = low elasticity; assumes a U.S. soybean supply elasticity of 0.22.

[c]HE = high elasticity; assumes a U.S. soybean supply elasticity of 0.92.

[d]Include all soybean producers.

[e]Included in consumers and producers.

SOURCE: Fernandez-Cornejo and Caswell, 2006; Qaim and Traxler, 2005.

agriculture (Feder et al., 1985) suggest that early adopters of new yield-increasing technologies gain early in the life of the technologies, but that their gains dissipate as prices go down. The United States was the dominant early adopter of GE varieties; James (2009) has since found a high rate of adoption of GE varieties more recently, mostly in developing countries. The agricultural products produced with genetic engineering are traded globally, and adoption of GE varieties worldwide affects prices that U.S. farmers receive.

According to James (2009), 331 million acres of land were planted to GE cultivars worldwide in 2009, of which nearly 95 percent was in six countries: the United States, Argentina, Brazil, Canada, India, and China. The total global acreage planted to GE crops in 2009 amounted to 8 percent of the world's tillable cropland. The GE cultivars were mostly of four crops: soybean (52 percent), corn (31 percent), cotton (12 percent), and canola (5 percent). In 2009, 77 percent of the soybean area, 49 percent of the cotton, 26 percent of the corn, and 21 percent of the canola lands were grown with GE cultivars. Much of the adoption of GE corn and cotton has been in the United States, Argentina, and Brazil, but China and India are major adopters of Bt cotton. The majority of acres planted to GE crops were HR varieties, at approximately 62 percent, followed by stacked traits at 21 percent and IR varieties at 15 percent. Stacked traits grew at a 23-percent rate from 2007 to 2008, the highest rate of the three trait categories (James, 2009).

According to Sexton et al. (2009), the high increases in yield that resulted from adoption of Bt cotton in developing countries have con-tributed to the increase in the world cotton supply and to the relatively low prices of cotton from 1998–2008. They suggest that the decline in the price of cotton relative to the price of other agricultural commodities has contributed to the transition from cotton to other crops in California. The same shift away from cotton is taking place in other cotton-producing regions. Total upland cotton-planted acreage in the United States has declined by 36.8 percent since 2002 (USDA-NASS, 2009a, 2009b).

Soybean acreage began to increase in the United States in 1997 and stayed relatively high until 2002, in part because the commodity support prices in the Federal Agricultural Improvement and Reform Act of 1996 favored soybean over other program crops. Even though Sexton et al. (2009) found that average yield of soybean—the crop with the highest rate of adoption of GE cultivars—grew more slowly than that of cotton after the introduction of GE varieties, the introduction of GE soybean contributed to the expansion of harvested soybean area worldwide, which grew by nearly 30 percent from 1997 to 2007 (FAO, 2008). In Argentina alone, GE soybean enabled adoption of no-till practices, which facilitated double cropping of wheat and soybean and contributed to a 9.9-million-acre increase in the soybean area from 1996 to 2003 (Trigo and Cap, 2003). The adoption of GE soybean in South America contributed to the increase in soybean supply, which also occurred because of the expansion of soybean acreage in Brazil. That supply shift caused downward pressure in soybean prices and had an adverse effect on growers in the United States, although the price effect was overwhelmed by the effect of increased global demand for soybean during the period 2006–2008.

Many of the analyses summarized in Tables 3-4 and 3-5 are based on partial-equilibrium models (in which the price of one good is examined and all other prices are held constant), but several studies have examined the effect of adoption of GE cultivars on producers and consumers by using a computable general-equilibrium approach (the prices of good are examined in relationship to one another). Some of the studies also attempted to assess the costs of access barriers imposed on GE crops by the European Union (EU), which has had a moratorium on the production and import of GE crops since 1999. Qaim (2009) has surveyed those studies and found that they predict an annual global welfare gain to consumers and producers from adoption of GE cultivars without restrictions, ranging from $1.4 billion from the adoption of Bt cotton to $10 billion from the adoption of GE oilseeds and corn. The results of the studies suggest that bans on imports of GE crops reduce the potential economic welfare of several parties, including U.S. farmers, but that European consumers suffer much of the loss.

Anderson and Jackson (2003) estimated that even under free trade— with global welfare gain from the introduction of GE cultivars of cotton, corn, and oilseeds that will enhance supply—farmers in the exporting countries will actually lose 0.07 percent of their income because of lower prices, whereas low-income consumers in North America will stand to gain from the introduction of GE cultivars because of lower food prices, all other things being equal. A moratorium on the export of GE crops to the EU will quadruple the losses to U.S. farmers. Such asynchronous conditions, when GE crops are approved at different times or not at all by different countries, could influence farmers' planting decisions because of those losses. Yet at the same time that U.S. farmers suffer economically, U.S. consumers benefit. A more severe moratorium on GE exports by the EU and other developed economies, such as that of Japan, is estimated to reduce the income of North American farmers by 0.5 percent. Those bans will hurt European consumers but benefit European farmers. Nielsen and Anderson (2001) showed that the welfare costs of less drastic barriers, such as labeling and segregation requirements, are important but smaller than the cost of bans.

Lence and collaborators (Lence and Hayes, 2005a, 2005b, 2006; Lence et al., 2005) showed that, in addition to the cost saving and other benefits, the overall welfare impact of genetic-engineering technology depends on the level of consumer concern with the technology and the costs of identity preservation. In particular, they state that their results suggest "the United States may have maximized welfare by not requiring labeling" of GE corn and soybean, but they claim that their results also suggest that "recently approved EU legislation enforcing labeling of GE crops also makes sense because consumer concern in the EU appears to be greater than that in the United States."[13]

The literature suggests that adoption of GE cultivars puts downward pressure on crop prices and increases the earnings of adopting farmers in the early years of the adoption process and that barriers to access reduce grower income. But there is a paucity of studies of the welfare effects of genetic-engineering technology in recent years, when adoption has increased globally, and this is an important subject for future research.

ECONOMIC IMPACTS ON OTHER PRODUCERS

Livestock Producers

Much of the soybean and corn produced in the United States is fed to livestock (Figures 3-6 and 3-7), and byproducts are used in consumer products, so quality and nutritional characteristics of soybean and corn associated with GE crops have been closely examined. Most studies of soybean have reported no differences in animal performance (Hammond et al., 1996); in important nutritional qualities, such as isoflavones (Duke et al., 2003); or in other characteristics at the macroscopic level of HR soybean (Magaña-Gómez and Calderón de la Barca, 2009). Researchers (Cox and Cherney, 2001; Jung and Sheaffer, 2004) have reported that glyphosate-resistant Bt corn does not affect feeding-quality characteristics of corn silage. Lutz et al. (2006) reported that the Bt protein Cry1Ab is degraded during the ensiling process. In feeding studies, there was no difference in milk production or milk composition between glyphosate-resistant corn, with or without the stacked Bt gene, and nontransgenic hybrids (Barrière et al., 2001; Phipps et al., 2005; Calsamiglia et al., 2007). There were no differences in body weight and feed use between rats fed grain from a Bt corn rootworm hybrid and rats fed grain from a nontransgenic hybrid (He et al., 2008). Likewise, no differences were observed in mortality, weight gain, feed efficiency, or carcass yield between broiler chickens fed grain from a Bt corn rootworm hybrid and chickens fed grain from a near-isoline (McNaughton et al., 2007). Thus, empirical studies have clearly

---

[13]The welfare implications of different regimes of protection of intellectual-property rights in the seed industry have also been studied (Lence et al., 2005).

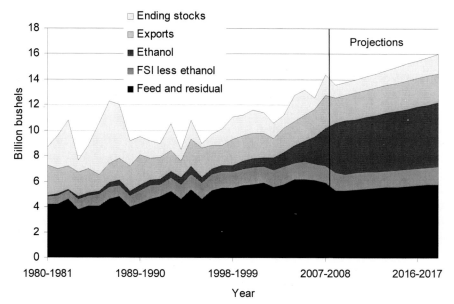

FIGURE 3-6 U.S. corn use.
NOTE: FSI = food, seed, and industrial.
SOURCE: USDA-ERS, 2009.

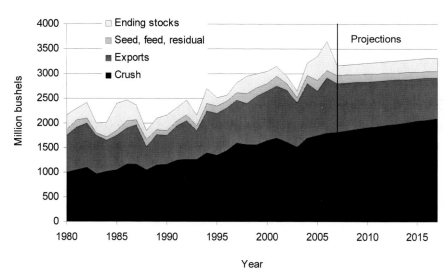

FIGURE 3-7 U.S. soybean use.
NOTE: Crush is used primarily for livestock feed.
SOURCE: USDA-ERS, 2009.

indicated that there is no adverse effect on quality of livestock feed or on the output or quality of livestock products.

Furthermore, nutritional characteristics of GE and conventional corn hybrids—including fatty acid profiles, mineral and vitamin contents, lutein, and total phenol and antioxidant activity—were comparable (Venneria et al., 2008) although some slight differences in triglycerides and urinary phosphorus and sodium extractions were noted in male rats (Magaña-Gómez and Calderón de la Barca, 2009). Cotton seed is used as a byproduct in animal feed, and cottonseed oil is used for human consumption. Castillo et al. (2004) found that Bt cotton seeds were deemed nutritionally equivalent with no difference in feed intake, milk yield, or composition. Few studies have been conducted to assess the levels of the glyphosate metabolite aminomethylphosphonic acid (AMPA) in glyphosate-treated, glyphosate-resistant corn hybrids; however, one study by Reddy et al. (2008) reported no detection of AMPA. Duke et al. (2003) reported that AMPA was detected in glyphosate-treated, glyphosate-resistant soybean seeds; however, EPA has not established a tolerance for AMPA. Given that AMPA is not considered significantly toxic (Giesy et al., 2000), the discovery of AMPA in glyphosate-treated, glyphosate-resistant soybean is not considered to be an issue of importance at this time.

Feed costs constitute nearly half the variable costs of livestock production, so even moderate price fluctuations can seriously affect the trajectory of the livestock market (USDA-NASS, 2008). As mentioned above, livestock operators are the buyers of feed, and they are the major beneficiaries of reductions in the prices of corn and soybean, to which the adoption of GE crops has contributed. They also benefit from increased feed safety from the reduction of mycotoxins (Wu, 2006). We are not aware of any quantitative estimation of savings to livestock operators and final consumers due to the adoption of GE crops or of the resulting effect on the profitability of livestock operations. This is another subject on which future research is desirable.

Producers of Non-Genetically Engineered Crops

The adoption of GE crops affects production costs for non-GE farmers in several key ways. GE crops alter the demand for inputs, and this affects the cost of inputs to GE and non-GE crops alike.[14] For example, Bt crop varieties that reduce insecticide use also lower the input costs for

---

[14]As stated above, the introduction of GE crops will probably reduce pest damage and, in some cases, will reduce the commodity prices of corn, soybean, and cotton. In the damage-control framework, the demand for inputs other than the ones controlling pests (such as water and energy) is represented by (Lichtenberg and Zilberman, 1986)

producers who use insecticides that substitute for Bt because the lower overall demand for them puts downward pressure on their prices. In other cases, GE crops increase the demand for other inputs. HR varieties increase demand for broad-spectrum herbicides, like glyphosate, which can have mixed effects on the price. On the one hand, the increase in demand puts upward pressure on the prices of those herbicides and, everything else being equal, increases the profits of the firms that manufacture GE seeds. On the other hand, the expanded market for broad-spectrum herbicides compatible with HR crops may allow firms to reduce the price of the herbicide but still increase profits through greater sales. HR varieties also affect the demand and prices for the herbicides that were used before HR crop varieties became available, usually by lowering prices because of reduced demand.

We have observed in Chapter 2 that GE crops can affect production of non-GE crops favorably or unfavorably through externalities associated with pest-control activities. To the extent that genetic-engineering technology successfully reduces pest pressure on a field, farmers of adjacent or nearby fields planted with non-GE crops may benefit from reductions in costs for pest control associated with reductions in regional pest populations (Sexton et al., 2007). Such favorable externalities may exist for Bt crops, which control pests that target GE and non-GE crops equally (Ando and Khanna, 2000). HR crops may provide some benefits to non-GE crops on adjacent fields by reducing rates of pollination of weeds, but more certain benefits will accrue to crops planted in rotation with GE crops. Specifically, because HR crops permit the postemergent use of broad-spectrum herbicides, such as glyphosate, weed species that affect GE and non-GE crops may be controlled more effectively. In particular, glyphosate has proved effective in controlling perennial weeds that appear late in the principal crop season, persist, and impose losses on subsequent crops (Padgette et al., 1996; Shaw and Arnold, 2002). The reduction in pest pressure from the late-season use of effective chemicals on HR crops

---

Quantity demanded = (crop price)(1 − damage)(marginal value of the input in producing potential output).

This equation suggests that GE cultivars will contribute to increased demand for inputs—such as fertilizer, water, and capital—if adoption of GE cultivars increases the earning per unit of potential output, which is equal to (crop price)(1 − damage), assuming no acreage constraints. Thus, when the introduction of GE cultivars does not affect crop prices but reduces pest damage, adoption of GE crops increases the demand for other input use. That increases the demands for fertilizer, water, capital, and so on, and causes upward pressure on their market prices. When the introduction of GE cultivars reduces commodity prices substantially, it may lead to reduced demand for other inputs. We are not familiar with empirical studies that have tried to estimate the impact of GE crops on the demand for or prices of other inputs, and this is a subject for future work.

may benefit the crop planted in the following season (Baylis, 2000; Tingle and Chandler, 2004) although empirical evidence of this effect is scarce. A massive field trial of crop rotation and herbicide application practices in Britain has provided evidence that the production systems used for HR canola can improve weed control in cereal crops planted in rotation (Sweet et al., 2004).

Farmers of non-GE crops may also experience adverse externalities associated with HR-crop weed control. Growers experience an adverse effect when an economically important amount of herbicide resistance builds up. As discussed in Chapter 2, resistance to broad-spectrum herbicides is a concern associated with adoption of HR varieties because use of other chemicals drastically declines in favor of the herbicide to which the crop is resistant (Shaner, 2000). When resistance in weeds evolves, farmers have resorted to managing those weeds with additional forms of control; they have either increased their use of the herbicide to which the HR crops are resistant, used additional and possibly more expensive forms of weed control (such as cultivation), or both. Such actions not only reduce or reverse the environmental benefits of HR crops reviewed in the previous chapter but also result in higher production costs for the grower compared to using glyphosate alone. To date, costs have not risen to the level of costs incurred in the conventional systems of weed control. If they had, a substantial reduction in the use of HR crops would have occurred. Resistance-management strategies, such as the use of refuges, can be expensive for individual farmers, though such strategies can provide long-run pest control benefits in the area that will offset the sum of individual costs if implemented correctly. Although Bt crops may be prone to resistance buildup because the toxins that target pests are always present in the field, the refuge requirements for Bt crops have thus far provided adequate protection from insect resistance buildup in the United States. The tradeoff is the requirement to plant some percentage of a crop to non-Bt cultivars, which may result in net economic costs to producers growing IR crops. Those costs, if they occur, are in the form of higher pesticide costs, foregone yield, or both. A benefit is the lower cost of seed for the refuge acres. A case in point is Bt cotton with the single trait for bollworm and budworm control, for which EPA requires a 20-percent refuge that can be treated with synthetic insecticides or a 5-percent refuge that cannot be treated with insecticides in the Southeast. Farmers who choose the 20-percent refuge can incur higher insecticide costs to treat insect infestations—more passes over the field and more labor to scout for insects—but have lower overall seed costs. Those who choose the 5-percent, untreated refuge can experience substantial yield loss on the refuge acres, though the cost of seed for those acres is lower. It is important to note that before the introduction of the Bt crops substantial insect resistance to other classes of insecticides, such as pyrethroids, had been observed.

SOCIOECONOMIC IMPACTS OF GENE FLOW

Inadvertent gene flow from GE to non-GE crops can also have a variety of social and economic effects. Both the Ecological Society of America and the National Research Council have recognized that some degree of gene flow between sexually compatible GE and non-GE crops occurs regularly (NRC, 2004; Snow et al., 2005). Indeed, the presence of adventitious GE traits in the intended non-GE seed supply of canola, cotton, corn, and soybean and in the seed supply of GE crops (e.g., a Bt trait in crop seed that is intended as HR only) is well documented (Beckie et al., 2003; Friesen et al., 2003; Mellon and Rissler, 2004; Heuberger et al., 2008; Heuberger and Carrière, 2009). The probability of gene flow is similar in both directions between GE and non-GE varieties of a crop (Mallory-Smith and Zapiola, 2008); however, for farmers, consumers, and food distributors, the actual and perceived consequences of gene flow from GE to non-GE crops are greater than the consequences of gene flow from non-GE to GE crops.

Gene flow between GE and non-GE crops occurs via three routes: cross-pollination between GE and non-GE plants from different fields (as discussed in "Gene Flow and Genetically Engineered Crops" in Chapter 2), co-mingling of seed before the production year (in the presence of GE traits in seed bags of non-GE crops) or during the production year (mixing of seed at planting, at harvest, or during storage), and germination of seeds left behind (i.e., volunteers) after the production year (Owen, 2005; McHughen, 2006). Generally, GE and non-GE crops can coexist. However, given that some domestic and foreign consumers are willing to pay a premium for non-GE products, there are strong market incentives as well as some sociocultural reasons for farmers, seed distributors, and food processors to minimize the adventitious presence of GE traits in non-GE crops and derived products (Lin et al., 2003; Belcher et al., 2005; Furtan et al., 2007; Devos et al., 2008).

Gene flow between HR and non-HR crops can increase production costs if gene flow promotes weediness. For example, when volunteer seeds survive and germinate to the following season, field management costs increase because the volunteers will not be eliminated by glyphosate applications. Similarly, if HR traits cross into weedy relatives, weed-control expenses will be higher for all fields on to which these weeds spread, whether the farmer grows GE crops or not (Smyth et al., 2002).

Gene flow of GE traits could jeopardize the economic value of the entire harvest of non–GE-crop farmers by rendering their output unsuitable for high-value markets (Bullock and Desquilbet, 2002). They could also have unfavorable effects on the levels of trust that exist between market participants. Two groups of farmers could be impacted by gene flow: those farming non-GE crops conventionally and organic farmers. The U.S. government does not have thresholds for what level of purity is required to characterize a product as non-GE; the thresholds are instead determined by the market. The U.S. National Organic Program excludes GE methods from the organic process (Organic Foods Production Act of 1990). Because of adventitious gene flow, the organic process does not necessarily result in a non-GE product when it goes to market; whether adventitious presence is discovered depends on if testing for GE material is conducted. Therefore, if GE traits are discovered in organic crops intended for a non-GE market, the organic or non-GE status of a crop may be forfeited depending on the potential legal or market tolerances for the presence of GE traits (Gealy et al., 2007). Other governments have set thresholds for organic and non-GE crops; for example, Japan has a 5-percent threshold for corn while the EU has zero tolerance for non-approved GE material but a 0.9-percent permissible level for GE material that has been approved by the EU (Bradford, 2006; Ronald and Fouche, 2006). Tests can be performed to assess the presence of GE traits in grain to preserve the identity of non-GE grain; whether a positive test results in rejection of a product depends on the individual policies of buyers. Additional research is needed to determine the extent to which screening is used and its relationship to variation in consumer desires for purity in the food supply. Although non-GE products can lose market value because of the adventitious presence of GE material, the price of GE products is not affected by the adventitious presence of non-GE material. Accordingly, gene flow between GE and non-GE crops imposes costs primarily on consumers and producers of GE-free crops (Smyth et al., 2002; Belcher et al., 2005; Devos et al., 2008). Such a need to protect the market value of non-GE products probably contributed to the creation of GE-free zones in some regions of the United States and in the EU (Jank et al., 2006; Furtan et al., 2007). Widespread use of GE crops in the United States may have forced some corporations that were producing GE-free products to move their operations to countries where GE crops were less prevalent (Mellon and Rissler, 2004). Nevertheless, a survey published in 2004 suggested that 92 percent of U.S. organic growers who responded to the survey had not incurred any direct costs or suffered losses attributable to the adventitious presence of GE crops (Brookes and Barfoot, 2004). However, it must be noted that there have been considerable increases in the adoption of GE crops in the United States as well as growth in U.S. organic-crop production and the market for non-GE products since this survey was conducted, that so far GE traits have been incorporated into a small number of crops that have few near-relatives in U.S. agriculture, and that few studies have analyzed trends in the socioeconomic impacts of gene flow.

A zero tolerance for the presence of GE traits in non-GE crops is generally impossible to manage and is not technically or economically feasible. Pollen transfer between sexually-compatible GE and non-GE crops is difficult, if not impossible, to prevent, and segregation between GE and non-GE products may be accomplished more easily and economically when nonzero thresholds for the adventitious presence of GE material in non-GE end-user products and seed are established. The goal of the thresholds is to set acceptable limits for the presence of GE traits that have been deemed safe and approved for human consumption. Accordingly, programs aimed at establishing such thresholds are analogous to seed-certification and food-labeling programs that have been used for decades to ensure the quality of seeds for agriculture and of food for consumers. The difficulty of maintaining the coexistence of GE and non-GE crops increases as the tolerance for the adventitious presence of GE traits in non-GE products becomes lower and the adventitious presence of GE traits in non-GE products becomes easier to detect even at very low levels due to technological advances.

The situation has a drastically greater impact when GE traits *not* approved for human consumption contaminate non-GE products. Such contamination can have strong adverse effects on market value, on the possibility of exporting crops, on the costs of remedial actions to remove contaminated supplies, and therefore on the profit margins of food producers and distributors (Lin et al., 2003; Vermij, 2006; Vogel, 2006). It can also undermine public confidence in the food system. The effects of the identification of a variety of Bt corn marketed under the name StarLink® in human food constitute an important example. StarLink® was approved only for use in animal feed but was discovered in products destined for human consumption. The resulting concerns about food safety led to the recall of more than 300 food products, and some major U.S. export markets, such as Japan and South Korea, imposed trade restrictions (Lin et al., 2003). The technology developer ultimately discontinued sale of StarLink® seed. Similarly, the accidental release of glufosinate-resistant rice in the United States in 2006 and the contamination of sulfonylurea-resistant flax in Canadian exports in 2009 imposed heavy costs on farmers, commodity traders, and processors.

Those examples of accidental releases, or any other low-level presence of unapproved GE material in the food supply, impose considerable costs on the food system that need to be accounted for in cost-benefit analyses of GE crops (Salazar et al., 2006). They also affect farmers' planting decisions because of the risk of lost revenues and other economic and social costs. As more exotic GE crops (e.g., pharmaceuticals) enter the commercialization phase, possible supply disruptions will multiply with greater potential for conflict between sectors in the food and nonfood industries and substantial economic costs. Such potential market and political repercussions indicate that a very low tolerance threshold set by U.S. regulatory authorities is appropriate for the presence of unapproved GE traits in food intended for human consumption.

Certain groups of consumers prefer GE-free products, a preference that is likely to increase the demand for products made with ingredients from organically grown crops. The NOP regulations require the certified organic producers must produce and handle their organic agricultural products without the use of GE methods (National Organic Program). However, the unintentional presence of GE material in organic products will not necessarily lead certifying agents to change the status of an organic product or operation (USDA-AMS, 2000). As explained above, because some level of gene flow between GE and non-GE crops is difficult to prevent, the adventitious presence of GE material has been detected in non-GE products, including certified organic products. Therefore, the process-based NOP standard that excludes GE methods in production and handling systems does not assure that organically grown crops with non-GE methods will be free of GE material for marketing.

The presence of GE material can affect the ability of growers to sell non-GE and organic crops in domestic and foreign markets with requirements beyond the process-based standard of the NOP. Accordingly, policies have been established to manage the potential for adventitious presence while enabling coexistence of GE, GE-free, and organic production systems. However, policy-established tolerance thresholds for the adventitious presences of traits from commercialized GE crops in non-GE or organic products vary considerably among countries. For example, in the United States, voluntary labeling of food as GE-free is allowed as long as a product contains less than 5-percent adventitious presence of GE material (Demont and Devos, 2008; Organic Foods Production Act of 1990). In contrast, the EU allows up to 0.9-percent adventitious GE material in non-GE food, animal feed, and products labeled as organic if the GE crop has been approved in the EU; otherwise, the threshold is zero (Demont and Devos, 2008). Certified non-GE seed sold to farmers in the United States is typically expected to contain less than 0.5-1 percent of seeds (depending on crop type) with GE traits (Mellon and Rissler, 2004; CCIA, 2007). Thresholds for commercial seed have been considered but have not yet been implemented uniformly in the EU (Kalaitzandonakes and Magnier, 2004; Devos et al., 2008).

GE-free or organic products lose their premium market value when the adventitious presence of GE material exceeds established government or market thresholds. Anecdotal stories suggest that the crops of U.S. organic growers are being screened in the marketing chain for the presence of GE material and are being rejected if levels exceed market-determined levels. We do not have evidence to judge how widespread such testing is in the United States. This issue deserves more investigation to determine the extent of such market-led behavior and the social and other factors driving it. We do know that given the threshold criteria in the EU for GE material in organic products, food produced in the United States and labeled as organic by U.S. certifiers could be rejected in the EU as not organic because of adventitious presence of GE material even though no GE seed or crops were used in production by U.S. producers. The coexistence of GE and non-GE products is possible as long as measures are taken to ensure that the adventitious presence of GE traits remains below the thresholds set in receiving markets, either by governments or buyers. In general, threshold differences among regions contribute to creating barriers to the use of GE crops and trade in non-GE products (Smyth et al., 2002; Demont and Devos, 2008; Devos et al., 2008).

Separating GE and non-GE products at every step of the production process is expensive, and costs increase as thresholds for the presence of GE traits in non-GE products decrease (Lin et al., 2003; Kalaitzandonakes and Magnier, 2004). Growers must attend to details and apply considerable effort to achieve effective segregation between GE and non-GE crops (CBI, 2007). Grain segregation in normal production settings is difficult but can be accomplished and could effectively minimize the co-mingling of GE and non-GE crops. Given that co-mingling of seeds can be costly for growers, particularly for growers who have specific contracts that restrict GE traits, tactics for isolating GE crops from non-GE crops must be established effectively (Owen, 2000). Controlling volunteer GE crops in non-GE crops may not be difficult, depending on crop rotation but requires considerable diligence on the part of growers (Owen, 2005). When volunteer crops acquire a GE trait for herbicide resistance via unintended gene flow, weed-management costs for a grower may increase and potential crop yield may decline if the crop planted the following season is also resistant to glyphosate (Owen and Zelaya, 2005). Furthermore, the isolation distances required to maintain complete segregation for open-pollinated crops are often too large to be economically feasible (Matus-Cádiz et al., 2004).

An economic assessment based on data from major seed firms in the United States indicated that reducing the adventitious presence of GE traits in non-GE corn seed from 1 percent to 0.3 percent would raise seed production costs by about 35 percent (Kalaitzandonakes and Magnier,

2004).[15] The increased costs would involve changes in field operations and in processing and result from new expenses for extra purity testing, storage, and transportation, but most of the increase in production costs would result from measures taken at the field level to minimize gene flow. Thus, programs that set levels of tolerance for the adventitious presence of GE traits in non-GE products probably have substantial impacts on growers directly and would increase the cost of non-GE seed and the market value of GE-free and organic products (Smyth et al., 2002; Kalaitzandonakes and Magnier, 2004; Belcher et al., 2005).

Barring the risk of contamination, GE crops can contribute to the creation of market opportunities for non-GE farmers. The organic market is a primary example. By virtue of the ban on the use of GE traits in the official USDA definition of organic production, the organic movement can market itself to, and collect a price premium from, consumers who prefer not to purchase food or fiber produced with genetic-engineering technology. Consumer preference for non-GE foods may be related to other traits associated with organic production, but the stated price premium for non-GE crops is substantial in some segments of the population (Huffman et al., 2003).

CONCLUSIONS

The widespread adoption of GE crops has had agronomic and economic implications for adopters and non-GE producers in the United States. For GE farmers, the general increase in yield, reduction in some input costs, improvement in pest control, increase in personal safety, and time-management benefits have generally outweighed the additional costs of GE seed. The use of HR crops has not greatly increased yields, but it has generally improved weed control, especially on farms where substantial weed resistance to the specific herbicide to which the HR crop is resistant has not developed, and it has improved farmers' incomes by saving time thus facilitating more off-farm work or providing more management time on the farm. IR crops have increased yields in areas where economically damaging insect-pest pressures occur and have saved on expenditures for conventional pesticide. Thus, the use of HR and IR crops has mostly increased adopters' incomes compared with the use of non-GE varieties.

It should be noted that the economic benefits have changed over time and probably will continue to change. Yield lag and yield drag were not

---

[15]This study is a summary assessment over various GE-crop technologies and therefore should not be applied to specific situations. It is likely that the impacts would vary considerably over different GE cultivars and their specific farming situations.

uncommon when HR crop varieties were first introduced, but GE traits have since been incorporated into high-yielding varieties, and improved GE events have replaced the initial events. Although research has identified those changes in farmers' experience with GE crops, there has been little investigation of the economic impact of GE crops more recently. More research would improve the information available to farmers, plant breeders, and policy makers as market, environmental, and social conditions change.

The extent to which GE crops make it economical to expand production to lands not previously cultivated or to intensify production on existing cropland with double cropping has not been reported adequately in the literature. More research on the economic effects of GE-crop adoption on non–GE-crop producers would also be beneficial. Examples include the costs and benefits of shifts in pest management for non-GE producers due to the adoption of GE crops, the value of market opportunities afforded to organic farmers by defining their products as non-GE crops, the economic impacts of GE adoption on livestock producers, and the costs to farmers, marketers, and processors of adventitious presence or contamination from approved or unapproved GE traits and crops into restricted markets.

REFERENCES

Ahmad, A., G.E. Wilde, R.J. Whitworth, and G. Zolnerowich. 2006. Effect of corn hybrids expressing the coleopteran-specific Cry3Bb1 protein for corn rootworm control on aboveground insect predators. *Journal of Economic Entomology* 99(4):1085–1095.

Al-Deeb, M.A., and G.E. Wilde. 2005. Effect of Bt corn expressing the Cry3Bb1 toxin on western corn rootworm (Coleoptera: Chrysomelidae) biology. *Journal of the Kansas Entomological Society* 78(2):142–152.

Alston, J.M., J.A. Hyde, M.C. Marra, and P.D. Mitchell. 2002. An ex ante analysis of the benefits from the adoption of corn rootworm resistant transgenic corn technology. *AgBioForum* 5(3):71–84.

Anderson, K., and L.A. Jackson. 2003. Why are US and EU policies toward GMOs so different? *AgBioForum* 6(3):95–100.

———. 2005. Global responses to GM food technology: Implications for Australia. RIRDC Publication No. 05/016. Canberra, ACT: Rural Industries Research and Development Corporation. Available online at https://rirdc.infoservices.com/au/downloads/05-016.pdf. Accessed March 31, 2010.

Ando, A.W., and M. Khanna. 2000. Environmental costs and benefits of genetically modified crops: Implications for regulatory strategies. *American Behavioral Scientist* (3):435–463.

Baker, C.J., K.E. Saxton, and W.R. Ritchie. 1996. *No–tillage seeding: Science and practice.* Oxford, UK: CABI Publishing.

Baker, C.J., K.E. Saxton, W.R. Ritchie, W.C.T. Chamen, D.C. Reicosky, F. Ribeiro, S.E. Justice, and P.R. Hobbs. 2007. *No-tillage seeding in conservation agriculture.* 2nd ed. Oxford, UK: CABI Publishing/UN-FAO.

Barrière, Y., R. Vérité, P. Brunschwig, F. Surault, and J.C. Emile. 2001. Feeding value of corn silage estimated with sheep and dairy cows is not altered by genetic incorporation of Bt176 resistance to *Ostrinia nubilalis*. *Journal of Dairy Science* 84(8):1863–1871.

Baute, T.S., M.K. Sears, and A.W. Schaafsma. 2002. Use of transgenic *Bacillus thuringiensis* Berliner corn hybrids to determine the direct economic impact of the European corn borer (Lepidoptera: Crambidae) on field corn in eastern Canada. *Journal of Economic Entomology* 95(1):57–64.

Baylis, A.D. 2000. Why glyphosate is a global herbicide: Strengths, weaknesses and prospects. *Pest Management Science* 56(4):299–308.

Beckie, H.J., S.I. Warwick, H. Nair, and G. Séguin-Swartz. 2003. Gene flow in commercial fields of herbicide-resistant canola (*Brassica napus*). *Ecological Applications* 13(5):1276–1294.

Belcher, K., J. Nolan, and P.W.B. Phillips. 2005. Genetically modified crops and agricultural landscapes: Spatial patterns of contamination. *Ecological Economics* 53(3):387–401.

Benbrook, C. 2004. The impact of genetically engineered crops on pesticide use: The first nine years. October. Technical Paper No. 7. Ag BioTech InfoNet. Sandpoint, ID. Available online at http://www.organic-center.org/reportfiles/Full_first_nine.pdf. Accessed April 8, 2009.

Bernard, J.C., J.D. Pesek, Jr., and C. Fan. 2004. Performance results and characteristics of adopters of genetically engineered soybeans in Delaware. *Agricultural and Resource Economics Review* 33(2):282–292.

Bertram, M.G., and P. Pedersen. 2004. Adjusting management practices using glyphosate-resistant soybean cultivars. *Agronomy Journal* 96(2):462–468.

Boyle, K.P. 2006. The economics of on-site conservation tillage. West National Technology Support Center technical note. September. Econ 101.01. U.S. Department of Agriculture–Natural Resources Conservation Service. Portland, OR. Available online at ftp://ftp-fc.sc.egov. usda.gov/Economics/Technotes/ConservationTill_01.doc. Accessed August 2, 2009.

Bradford, K.J. 2006. Methods to maintain genetic purity of seed stocks. *Agricultural biotechnology in California*. University of California–Division of Agriculture and Natural Resources. Publication 8189. Available online at http://ucbiotech.org/resources/ factsheets/8189.pdf. Accessed March 31, 2010.

Bradley, K.W., and L.E. Sweets. 2008. Influence of glyphosate and fungicide coapplications on weed control, spray penetration, soybean response, and yield in glyphosate-resistant soybean. *Agronomy Journal* 100(5):1360–1365.

Bradley, K.W., N.H. Monnig, T.R. Legleiter, and J.D. Wait. 2007. Influence of glyphosate tank-mix combinations and application timings on weed control and yield in glyphosate-resistant soybean. *Crop Management*. Available online at http://www.plantmanagementnetwork. org/pub/cm/research/2007/tank/. Accessed April 7, 2009.

Brookes, G., and P. Barfoot. 2004. Co-existence in North American agriculture: Can GM crops be grown with conventional and organic crops? PG Economics Ltd. Dorchester, UK. Available online at http://www.pgeconomics.co.uk/pdf/CoexistencereportNAmericafinalJune2004. pdf. Accessed May 15, 2009.

Bullock, D.S., and M. Desquilbet. 2002. The economics of non-GMO segregation and identity preservation. *Food Policy* 27(1):81–99.

Bullock, D.S., and E.I. Nitsi. 2001. Roundup Ready soybean technology and farm production costs: Measuring the incentive to adopt genetically modified seeds. *American Behavioral Scientist* (8):1283–1301.

Calsamiglia, S., B. Hernandez, G.F. Hartnell, and R. Phipps. 2007. Effects of corn silage derived from a genetically modified variety containing two transgenes on feed intake, milk production, and composition, and the absence of detectable transgenic deoxyribonucleic acid in milk in Holstein dairy cows. *Journal of Dairy Science* 90(10):4718–4723.

Carpenter, J., L.L. Wolfenbarger, and P.R. Phifer. 2001. GM crops and patterns of pesticide use. *Science* 292(5517):637b–638.

Carpenter, J., A. Felsot, T. Goode, M. Hammig, D. Onstad, and S. Sankula. 2002. *Comparative environmental impacts of biotechnology-derived and traditional soybean, corn, and cotton crops.* Ames, IA: Council for Agricultural Science and Technology. Available online a http://www.soyconnection.com/soybean_oil/pdf/EnvironmentalImpactStudy-English.pdf. Accessed March 31, 2010.

Carrière, Y., C. Ellers-Kirk, M.S. Sisterson, L. Antilla, M. Whitlow, T.J. Dennehy, and B.E. Tabashnik. 2003. Long-term regional suppression of pink bollworm by *Bacillus thuringiensis* cotton. *Proceedings of the National Academy of Sciences of the United States of America* 100(4):1519–1523.

Carrière, Y., M.S. Sisterson, and B.E. Tabashnik. 2004. Resistance management for sustainable use of *Bacillus thuringiensis* crops in integrated pest management. In *Insect pest management: Field and protected crops.* eds. A.R. Horowitz and I. Ishaaya, pp. 65–95. Berlin: Springer.

Castillo, A.R., M.R. Gallardo, M. Maciel, J.M. Giordano, G.A. Conti, M.C. Gaggiotti, O. Quaino, C. Gianni, and G.F. Hartnell. 2004. Effects of feeding rations with genetically modified whole cottonseed to lactating Holstein cows. *Journal of Dairy Science* 87(6):1778–1785.

Catangui, M.A., and R.K. Berg. 2002. Comparison of *Bacillus thuringiensis* corn hybrids and insecticide-treated isolines exposed to bivoltine European corn borer (Lepidoptera: Crambidae) in South Dakota. *Journal of Economic Entomology* 95(1):155–166.

Cattaneo, M.G., C.M. Yafuso, C. Schmidt, C.-Y. Huang, M. Rahman, C. Olson, C. Ellers-Kirk, B.J. Orr, S.E. Marsh, L. Antilla, P. Dutilleul, and Y. Carrière. 2006. Farm-scale evaluation of the impacts of transgenic cotton on biodiversity, pesticide use, and yield. *Proceedings of the National Academy of Sciences of the United States of America* 103(20):7571–7576.

CBI (Council for Biotechnology Information). 2007. *Can biotech and organic crops coexist?* Washington, DC: Council for Biotechnology Information.

CCIA (California Crop Improvement Association). 2007. Certification standards. Davis: Parsons Seed Certification Center/California Crop Improvement Association. Available online at http://ccia.ucdavis.edu/. Accessed June 26, 2009.

Chase, C.A., and M.D. Duffy. 1991. An economic analysis of the Nashua tillage study: 1978–1987. *Journal of Production Agriculture* 4(1):91–98.

Clements, M.J., K.W. Campbell, C.M. Maragos, C. Pilcher, J.M. Headrick, J.K. Pataky, and D.G. White. 2003. Influence of Cry1Ab protein and hybrid genotype on fumonisin contamination and *Fusarium* ear rot of corn. *Crop Science* 43(4):1283–1293.

Corrigan, K.A., and R.G. Harvey. 2000. Glyphosate with and without residual herbicides in no-till glyphosate-resistant soybean (*Glycine max*). *Weed Technology* 14(3):569–577.

Couvillion, W.C., F. Kari, D. Hudson, and A. Allen. 2000. A preliminary economic assessment of Roundup Ready soybeans in Mississippi. May Research Report 2000–005. Mississippi State University. Starkville, MS. Available online at http://ageconsearch. umn.edu/bitstream/15783/1/rr00-005.pdf. Accessed April 28, 2009.

Cox, W.J., and D.J.R. Cherney. 2001. Influence of brown midrib, leafy, and transgenic hybrids on corn forage production. *Agronomy Journal* 93(4):790–796.

Cox, W.J., R.R. Hahn, P.J. Stachowski, and J.H. Cherney. 2005. Weed interference and glyphosate timing affect corn forage yield and quality. *Agronomy Journal* 97(3):847–853.

Cox, W.J., R.R. Hahn, and P.J. Stachowski. 2006. Time of weed removal with glyphosate affects corn growth and yield components. *Agronomy Journal* 98(2):349–353.

Cox, W.J., J.H. Cherney, and E. Shields. 2007a. Clothianidin seed treatments inconsistently affect corn forage yield when following soybean. *Agronomy Journal* 99(2):543–548.

Cox, W.J., E. Shields, and J.H. Cherney. 2007b. The effect of clothianidin seed treatments on corn growth following soybean. *Crop Science* 47(6):2482–2485.

Cox, W.J., E. Shields, D.J.R. Cherney, and J.H. Cherney. 2007c. Seed-applied insecticides inconsistently affect corn forage in continuous corn. *Agronomy Journal* 99(6):1640–1644.

Cox, W.J., J. Hanchar, and E. Shields. 2009. Stacked corn hybrids show inconsistent yield and economic responses in New York. *Agronomy Journal* 101(6):1530–1537.

Culpepper, A.S., and A.C. York. 1999. Weed management and net returns with transgenic, herbicide-resistant, and nontransgenic cotton (*Gossypium hirsutum*). *Weed Technology* 13(2):411–420.

Dalley, C.D., J.J. Kells, and K.A. Renner. 2004. Effect of glyphosate application timing and row spacing on weed growth in corn (*Zea mays*) and soybean (*Glycine max*). *Weed Technology* 18(1):177–182.

Demont, M., and Y. Devos. 2008. Regulating coexistence of GM and non-GM crops without jeopardizing economic incentives. *Trends in Biotechnology* 26(7):353–358.

Devos, Y., M. Demont, and O. Sanvido. 2008. Coexistence in the EU—return of the moratorium on GM crops? *Nature Biotechnology* 26(11):1223–1225.

Dillehay, B.L., G.W. Roth, D.D. Calvin, R.J. Kratochvil, G.A. Kuldau, and J.A. Hyde. 2004. Performance of Bt corn hybrids, their near isolines, and leading corn hybrids in Pennsylvania and Maryland. *Agronomy Journal* 96(3):818–824.

Downs, H.W., and R.W. Hansen. 1998. Estimating farming fuel requirements. Farm & ranch series. No. 5.006. Colorado State University Extension Service. Fort Collins, CO. Available online at http://www.cde.state.co.us/artemis/ucsu20/ucsu2062250061998internet.pdf. Accessed May 4, 2009.

Duffy, M. 2001. Who benefits from biotechnology? Paper presented at the American Seed Trade Association Annual Meeting (Chicago, IL, December 5–7, 2001). American Seed Trade Association. Available online at http://www.econ.iastate.edu/faculty/duffy/Pages/biotechpaper.pdf. Accessed February 21, 2009.

Duke, S.O., A.M. Rimando, P.F. Pace, K.N. Reddy, and R.J. Smeda. 2003. Isoflavone, glyphosate, and aminomethylphosphonic acid levels in seeds of glyphosate-treated, glyphosate-resistant soybean. *Journal of Agricultural and Food Chemistry* 51(1):340–344.

Ellsworth, P.C., A. Fournier, and T.D. Smith. 2009. Arizona cotton insect losses. Publ. No. AZ1183. University of Arizona–College of Agriculture and Life Sciences. Tucson, AZ. Available online at http://cals.arizona.edu/crops/cotton/insects/cil/cil.html. Accessed April 20, 2009.

Elmore, R.W., F.W. Roeth, R.N. Klein, S.Z. Knezevic, A. Martin, L.A. Nelson, and C.A. Shapiro. 2001a. Glyphosate-resistant soybean cultivar response to glyphosate. *Agronomy Journal* 93(2):404–407.

Elmore, R.W., F.W. Roeth, L.A. Nelson, C.A. Shapiro, R.N. Klein, S.Z. Knezevic, and A. Martin. 2001b. Glyphosate-resistant soybean cultivar yields compared with sister lines. *Agronomy Journal* 93(2):408–412.

Falck-Zepeda, J.B., G. Traxler, and R.G. Nelson. 1999. Rent creation and distribution from the first three years of planting Bt cotton. ISAAA Briefs No. 14. The International Service for the Acquisition of Agri-biotech Applications. Ithaca, NY.

———. 2000a. Surplus distribution from the introduction of a biotechnology innovation. *American Journal of Agricultural Economics* 82(2):360–369.

———. 2000b. Rent creation and distribution from biotechnology innovations: The case of Bt cotton and herbicide-tolerant soybeans in 1997. *Agribusiness* 16(1):21–32.

FAO (Food and Agriculture Organization). 2008. FAOStat: ResourceSTAT. Available online at http://faostat.fao.org. Accessed June 16, 2009.

Feder, G., R.E. Just, and D. Zilberman. 1985. Adoption of agricultural innovations in developing countries: A survey. *Economic Development and Cultural Change* 33(2):255–298.

Fernandez-Cornejo, J. 2004. The seed industry in U.S. agriculture: An exploration of data and information on crop seed markets, regulation, industry structure, and research and development. Agriculture Information Bulletin No. 786. U.S. Department of Agriculture–Economic Research Service. Washington, DC. Available online at http://www.ers.usda.gov/publications/aib786/aib786.pdf. Accessed May 26, 2009.

Fernandez-Cornejo, J., and M.F. Caswell. 2006. The first decade of genetically engineered crops in the United States. Economic Information Bulletin No. 11. April. U.S. Department of Agriculture–Economic Research Service. Washington, D.C. Available online at http://www.ers.usda.gov/publications/eib11/eib11.pdf. Accessed June 15, 2009.

Fernandez-Cornejo, J., and A. Gregory. 2004. Managerial intensity and the adoption of conservation tillage. Paper presented at the Northeastern Agricultural and Resource Economics Association Annual Meeting (Halifax, Nova Scotia, June 20–23, 2004).

Fernandez-Cornejo, J., and C. Hendricks. 2003. Off-farm work and the economic impact of adopting herbicide-tolerant crops. Paper presented at the American Agricultural Economics Association Annual Meeting (Montreal, Canada, July 27–30, 2003). Available online at http://ageconsearch.umn.edu/bitstream/22130/1/sp03fe01/pdf. Accessed June 28, 2009.

Fernandez-Cornejo, J., and J. Li. 2005. The impacts of adopting genetically engineered crops in the USA: The case of Bt corn. Paper presented at the American Agricultural Economics Association Annual Meeting (Providence, RI, July 24–27, 2005). Available online at http://ageconsearch.umn.edu/bitstream/19318/1/sp05fe01.pdf. Accessed February 19, 2009.

Fernandez-Cornejo, J., and W.D. McBride. 2002. Adoption of bioengineered crops. Agricultural Economic Report No. 810. May 1. U.S. Department of Agriculture–Economic Research Service. Washington, DC. Available online at http://www.ers.usda.gov/publications/aer810/aer810.pdf. Accessed July 25, 2009.

Fernandez-Cornejo, J., and A.K. Mishra. 2007. Off-farm income, production decisions, and farm economic performance. ERR#36. U.S. Department of Agriculture–Economic Research Service. Washington, DC. Available online at http://www.ers.usda.gov/publications/err36/err36_reportsummary.pdf. Accessed June 29, 2009.

Fernandez-Cornejo, J., C. Klotz-Ingram, and S. Jans. 2002a. Farm-level effects of adopting herbicide-tolerant soybeans in the U.S.A. *Journal of Agricultural & Applied Economics* 34(1):149–163.

Fernandez-Cornejo, J., C. Alexander, and R.E. Goodhue. 2002b. Dynamic diffusion with disadoption: The case of crop biotechnology in the USA. *Agricultural and Resource Economics Review* 31(1):112–126.

Fernandez-Cornejo, J., C. Hendricks, and A.K. Mishra. 2005. Technology adoption and off-farm household income: The case of herbicide-tolerant soybeans. *Journal of Agricultural & Applied Economics* 37(3):549–563.

Fernandez-Cornejo, J., R. Lubowski, and A. Somwaru. 2007. Global adoption of agricultural biotechnology: Modeling and preliminary results. Paper presented at the 10th Annual Conference on Global Economic Analysis (West Lafayette, IN, June 7–9, 2007).

Fernandez-Cornejo, J., R. Nehring, E.N. Sinha, A. Grube, and A. Vialou. 2009. Assessing recent trends in pesticide use in U.S. agriculture. Paper presented at the 2009 Annual Meeting of the Agricultural and Applied Economics Association (Milwaukee, WI, July 26–28, 2009). Available online at http://agecomsearch.umn.edu/handle/49271. Accessed June 16, 2009.

Ferraro, P.J. 2009. Counterfactual thinking and impact evaluation in environmental policy. *New Directions for Evaluation* 2009(122):75–84.

Friesen, L.F., A.G. Nelson, and R.C. Van Acker. 2003. Evidence of contamination of pedigreed canola (*Brassica Napus*) seedlots in Western Canada with genetically engineered herbicide resistance traits. *Agronomy Journal* 95(5):1342–1347.

Frisvold, G., and M. Marra. 2004. The difficulty with data: How sampling and aggregation can affect measures of pesticide use in biotech crops. Paper presented at the 8th Annual International Consortium on Agricultural Biotechnology Research Conference (Ravello, Italy, July 8–11, 2004).

Frisvold, G.B., R. Tronstad, and J. Mortensen. 2000. Adoption of Bt cotton: Regional differences in producer costs and returns. In *2000 Proceedings Beltwide Cotton Conferences*, pp. 337–340 (San Antonio, TX, January 4–8, 2000). eds. P. Dugger and D. Richter. National Cotton Council of America.

Furtan, W.H., A. Güzel, and A.S. Weseen. 2007. Landscape clubs: Co-existence of genetically modified and organic crops. *Canadian Journal of Agricultural Economics* 55(2):185–195.

Gardner, J.G., and C.H. Nelson. 2007. Genetically modified crops and labor savings in US crop production. Paper presented at the 2007 Southern Agricultural Economics Association Annual Meeting (Mobile, AL, February 4–7, 2007).

Gealy, D.R., K.J. Bradford, L. Hall, R. Hellmich, A. Raybould, J. Wolt, and D. Zilberman. 2007. Implications of gene flow in the scale-up and commercial use of biotechnology-derived crops: Economic and policy considerations. Council for Agricultural Science and Technology, December (Issue 37). Available online at http://www.cast-science.org/websiteUploads/publicationPDFs/CAST%20Issue%20Paper%2037%20galley-final-2149.pdf. Accessed October 21, 2007.

Gianessi, L.P., and J.E. Carpenter. 1999. Agricultural biotechnology: Insect control benefits. National Center for Food and Agricultural Policy. Washington, DC. Available online at http://www.ncfap.org/documents/insectcontrolbenefits.pdf. Accessed May 20, 2009.

Giesy, J.P., S. Dobson, and K.R. Solomon. 2000. Ecotoxicological risk assessment for Roundup® herbicide. *Reviews of Environmental Contamination and Toxicology* 167:35–120.

Gower, S.A., M.M. Loux, J. Cardina, and S.K. Harrison. 2002. Effect of planting date, residual herbicide, and postemergence application timing on weed control and grain yield in glyphosate-tolerant corn (*Zea mays*). *Weed Technology* 16(3):488–494.

Gray, M.E., K.L. Steffey, R.E. Estes, and J.B. Schroeder. 2007. Responses of transgenic maize hybrids to variant western corn rootworm larval injury. *Journal of Applied Entomology* 131(6):386–390.

Griliches, Z. 1957. Hybrid corn: An exploration in the economics of technological change. *Econometrica* 25(4):501–522.

Hammond, B.G., J.L. Vicini, G.F. Hartnell, M.W. Naylor, C.D. Knight, E.H. Robinson, R.L. Fuchs, and S.R. Padgette. 1996. The feeding value of soybeans fed to rats, chickens, catfish and dairy cattle is not altered by genetic incorporation of glyphosate tolerance. *Journal of Nutrition* 126(3):717–727.

Harman, W.L., D.C. Hardin, A.F. Wiese, P.W. Unger, and J.T. Musick. 1985. No-till technology: Impacts on farm income, energy use and groundwater depletion in the plains. *Western Journal of Agricultural Economics* 10(1):134–146.

Harrington, J. 2006. University research finds Herculex RW twice as effective as YieldGard RW Against Corn Rootworm. *Crop Management*. Available online at http://www.plant-managementnetwork.org/pub/cm/news/2006/YieldGard/. Accessed April 7, 2009.

He, X.Y., K.L. Huang, X. Li, W. Qin, B. Delaney, and Y.B. Luo. 2008. Comparison of grain from corn rootworm resistant transgenic DAS-59122-7 maize with non-transgenic maize grain in a 90-day feeding study in Sprague-Dawley rats. *Food and Chemical Toxicology* 46(6):1994–2002.

Heimlich, R.E., J. Fernandez-Cornejo, W. McBride, C. Klotz-Ingram, S. Jans, and N. Brooks. 2000. Genetically engineered crops: Has adoption reduced pesticide use? *Agricultural Outlook* 273:13–17.

Heuberger, S., and Y. Carrière. 2009. Pollen-mediated transgene flow in agricultural seed production. Paper presented at the 94th Ecological Society of America annual meeting (Albuquerque, NM, August 2–7, 2009).

Heuberger, S., C. Yafuso, G. Degrandi-Hoffman, B.E. Tabashnik, Y. Carrière, and T.J. Dennehy. 2008. Outcrossed cottonseed and adventitious Bt plants in Arizona refuges. *Environmental Biosafety Research* 7(2):87–96.

Hubbell, B.J., M.C. Marra, and G.A. Carlson. 2000. Estimating the demand for a new technology: Bt cotton and insecticide policies. *American Journal of Agricultural Economics* 82(1):118–132.

Huffman, W.E., J.F. Shogren, M. Rousu, and A. Tegene. 2003. Consumer willingness to pay for genetically modified food labels in a market with diverse information: Evidence from experimental auctions. *Journal of Agricultural and Resource Economics* 28(3):481–502.

Huso, S.R., and W.W. Wilson. 2006. Producer surplus distributions in GM crops: The ignored impacts of Roundup Ready® wheat. *Journal of Agricultural and Resource Economics* 31(2):339–354.

Jackson, R.E., J.R. Bradley Jr., and J.W. Van Duyn. 2003. Field performance of transgenic cottons expressing one or two *Bacillus thuringiensis* endotoxins against bollworm, *Helicoverpa zea* (Boddie). *Journal of Cotton Science* 7(3):57–64.

James, C. 2009. *Global status of commercialized biotech/GM crops: 2009.* ISAAA Brief No. 41 ed. The International Service for the Acquisition of Agri-biotech Applications. Ithaca, NY.

Jank, B., J. Rath, and H. Gaugitsch. 2006. Co-existence of agricultural production systems. *Trends in Biotechnology* 24(5):198–200.

Jasa, P.J. 2000. Conservation tillage systems. Paper presented at the International Symposium on Conservation Tillage (Mazatlan, Mexico, January 24–27, 2000). Available online at http://agecon.okstate.edu/isct/labranza/jasa/tillagesys.doc. Accessed August 9, 2009.

Johnson, W.G., P.R. Bradley, S.E. Hart, M.L. Buesinger, and R.E. Massey. 2000. Efficacy and economics of weed management in glyphosate-resistant corn (*Zea mays*). *Weed Technology* 14(1):57–65.

Jung, H.G., and C.C. Sheaffer. 2004. Influence of Bt transgenes on cell wall lignification and digestibility of maize stover for silage. *Crop Science* 44(5):1781–1789.

Just, R.E., and D.L. Hueth. 1993. Multimarket exploitation: The case of biotechnology and chemicals. *American Journal Agricultural Economics* 75:936–945.

Kalaitzandonakes, N., and A. Magnier. 2004. Biotech labeling standards and compliance costs in seed production. *Choices: The magazine of food, farm and resource issues* 2nd Quarter:1–6. Available at http://www.choicesmagazine.org/2004-1/2004-2-01.pdf. Accessed March 31, 2010.

Kneževič, S.Z., S.P. Evans, and M. Mainz. 2003a. Row spacing influences the critical timing for weed removal in soybean (*Glycine max*). *Weed Technology* 17(4):666–673.

———. 2003b. Yield penalty due to delayed weed control in corn and soybean. *Crop Management.* Available online at http://www.plantmanagementnetwork.org/pub/cm/research/2003/delay/. Accessed April 8, 2009.

Krull, C.F., J.M. Prescott, and C.W. Crum. 1998. Seed marketing and distribution. In *Maize seed industries in developing countries.* ed. M.L. Morris, pp. 125–141. Boulder, CO: Lynne Rienner Publishers/CIMMYT.

Lauer, J. 2006. Performance of transgenic corn and soybean. Paper presented at the 2006 Annual ASA-CSSA-SSSA Meeting (Indianapolis, IN, November 12–16, 2006).

Lauer, J., and J. Wedberg. 1999. Grain yield of initial Bt corn hybrid introductions to farmers in the northern Corn Belt. *Journal of Production Agriculture* 12(3):373–376.

Lence, S.H., and D.J. Hayes. 2005a. Technology fees versus GURTs in the presence of spillovers: World welfare impacts. *AgBioForum* 8(2&3):172–186.

———. 2005b. Genetically modified crops: Their market and welfare impacts. *American Journal of Agricultural Economics* 87(4):931–950.

Lence, S.H., D.J. Hayes, A. McCunn, S. Smith, and W.S. Niebur. 2005. Welfare impacts of intellectual property protection in the seed industry. *American Journal of Agricultural Economics* 87(4):951–968.

Lence, S.H., and D.J. Hayes. 2006. EU and US regulations for handling and transporting genetically modified grains: Are both positions correct? *EuroChoices* 5(2):20–27.

Lichtenberg, E., and D. Zilberman. 1986. The econometrics of damage control—why specification matters. *American Journal of Agricultural Economics* 68(2):261–273.

Lin, W., G.K. Price, and E.W. Allen. 2003. StarLink: Impacts on the U.S. corn market and world trade. *Agribusiness* 19(4):473–488.

Lutz, B., S. Wiedemann, and C. Albrecht. 2006. Degradation of transgenic Cry1Ab DNA and protein in Bt-176 maize during the ensiling process. *Journal of Animal Physiology and Animal Nutrition* 90(3–4):116–123.

Ma, B.L., and K.D. Subedi. 2005. Development, yield, grain moisture and nitrogen uptake of Bt corn hybrids and their conventional near-isolines. *Field Crops Research* 93(2–3):199–211.

Ma, B.L., F. Meloche, and L. Wei. 2009. Agronomic assessment of Bt trait and seed or soil-applied insecticides on the control of corn rootworm and yield. *Field Crops Research* 111(3):189–196.

Ma, B.L., K. Subedi, L. Evenson, and G. Stewart. 2005. Evaluation of detection methods for genetically modified traits in genotypes resistant to European corn borer and herbicides. *Journal of Environmental Science and Health - Part B Pesticides, Food Contaminants, and Agricultural Wastes* 40(4):633–644.

Magaña-Gómez, J.A., and A.M. Calderón de la Barca. 2009. Risk assessment of genetically modified crops for nutrition and health. *Nutrition Reviews* 67(1):1–16.

Mallory-Smith, C., and M. Zapiola. 2008. Gene flow from glyphosate-resistant crops. *Pest Management Science* 64(4):428–440.

Marra, M.C. 2001. Agricultural biotechnology: A critical review of the impact evidence to date. In *The future of food: Biotechnology markets and policies in an international setting.* ed. P.G. Pardey, pp. 155–184. Washington, DC: International Food Policy Research Institute.

Marra, M.C., and N.E. Piggott. 2006. The value of non-pecuniary characteristics of crop biotechnologies: A new look at the evidence. In *Regulating agricultural biotechnology: Economics and policy.* eds. R.E. Just, J.M. Alston, and D. Zilberman, pp. 145–178. New York: Springer.

Marra, M.C., P.G. Pardey, and J.M. Alston. 2002. The payoffs to transgenic field crops: An assessment of the evidence. *AgBioForum* 5(2):43–50.

Marra, M.C., N.E. Piggott, and G.A. Carlson. 2004. The net benefits, including convenience, of Roundup Ready® soybeans: Results from a national survey. Technical Bulletin No. 2004-3. NSF Center for Integrated Pest Management. Raleigh, NC. Available online at http://cipm.ncsu.edu/cipmpubs/marra_soybeans.pdf. Accessed April 8, 2009.

Matus-Cádiz, M.A., P. Hucl, M.J. Horak, and L.K. Blomquist. 2004. Gene flow in wheat at the field scale. *Crop Science* 44(3):718–727.

May, O.L., and E.C. Murdock. 2002. Yield ranks of glyphosate-resistant cotton cultivars are unaffected by herbicide systems. *Agronomy Journal* 94(4):889–894.

May, O.L., A.S. Culpepper, R.E. Cerny, C.B. Coots, C.B. Corkern, J.T. Cothren, K.A. Croon, K.L. Ferreira, J.L. Hart, R.M. Hayes, S.A. Huber, A.B. Martens, W.B. McCloskey, M.E. Oppenhuizen, M.G. Patterson, D.B. Reynolds, Z.W. Shappley, J. Subramani, T.K. Witten, A.C. York, and B.G. Mullinix Jr. 2004. Transgenic cotton with improved resistance to glyphosate herbicide. *Crop Science* 44(1):234–240.

McHughen, A. 2006. The limited value of measuring gene flow via errant pollen from GM plants. *Environmental Biosafety Research* 5:1–2.

McNaughton, J.L., M. Roberts, D. Rice, B. Smith, M. Hinds, J. Schmidt, M. Locke, A. Bryant, T. Rood, R. Layton, I. Lamb, and B. Delaney. 2007. Feeding performance in broiler chickens fed diets containing DAS-59122-7 maize grain compared to diets containing non-transgenic maize grain. *Animal Feed Science and Technology* 132(3–4):227–239.

Mellon, M., and J. Rissler. 2004. *Gone to seed: Transgenic contaminants in the traditional seed supply.* Cambridge, MA: Union of Concerned Scientists.

Mitchell, J.P., D.S. Munk, B. Prys, K.M. Klonsky, J.F. Wroble, and R.L.D. Moura. 2006. Conservation tillage production systems compared in San Joaquin Valley cotton. *California Agriculture* 60(3):140–145.

Moschini, G., and H. Lapan. 1997. Intellectual property rights and the welfare effects of agricultural R&D. *American Journal of Agricultural Economics* 79(4):1229–1242.

Moschini, G., H. Lapan, and A. Sobolevsky. 2000. Roundup Ready® soybeans and welfare effects in the soybean complex. *Agribusiness* 16(1):33–55.

Mulugeta, D., and C.M. Boerboom. 2000. Critical time of weed removal in glyphosate-resistant *Glycine max*. *Weed Science* 48(1):35–42.

Myers, M.W., W.S. Curran, M.J. VanGessel, B.A. Majek, B.A. Scott, D.A. Mortensen, D.D. Calvin, H.D. Karsten, and G.W. Roth. 2005. The effect of weed density and application timing on weed control and corn grain yield. *Weed Technology* 19(1):102–107.

Naseem, A., and C. Pray. 2004. Economic impact analysis of genetically modified crops. In *Handbook of plant biotechnology*. eds. P. Christou and H.J. Klee, pp. 959–991. Hoboken, N.J.: John Wiley & Sons.

National Organic Program, Title 7 CFR 205.105.

Nielsen, C.P., and K. Anderson. 2001. Global market effects of alternative European responses to genetically modified organisms. *Weltwirtschaftliches Archiv* 137(2):320–346.

Nielsen, R.L. 2000. Transgenic crops in Indiana: Short-term issues for farmers. West Lafayette, IN: Purdue University. Available online at http://www.agry.purdue.edu/ext/corn/news/articles.00/GMO_Issues-000203.html. Accessed April 8, 2009.

NRC (National Research Council). 2004. *Biological confinement of genetically engineered organisms*. Washington, DC: National Academies Press.

Oplinger, E.S., M.J. Martinka, and K.A. Schmitz. 1998. Performance of transgenic soybeans: Northern United States. In *Proceedings of the 28th soybean seed research conference*, pp. 10–14 (Chicago, IL, December 1998). Alexandria, VA: American Seed Trade Association.

Organic Foods Production Act of 1990 7 U.S.C. sec 6501 et seq.

Owen, M.D.K. 2000. Current use of transgenic herbicide-resistant soybean and corn in the USA. *Crop Protection* 19(8–10):765–771.

Owen, M.D.K. 2005. Maize and soybeans—controllable volunteerism without ferality? In *Crop ferality and volunteerism*. ed. J. Gressel, pp. 149–165. Boca Raton, FL: Taylor & Francis.

Owen, M.D.K., and I.A. Zelaya. 2005. Herbicide-resistant crops and weed resistance to herbicides. *Pest Management Science* 61(3):301–311.

Padgette, S.R., D.B. Re, G.F. Barry, D.E. Eichholz, X. DeLannay, R.L. Fuchs, G. Kishore, and R.T. Fraley. 1996. New weed control opportunities: Development of soybeans with a Roundup Ready™ Gene. In *Herbicide-resistant crops: Agricultural, environmental, economic, regulatory, and technical aspects*. ed. S.O. Duke, pp. 53–84. Boca Raton: Lewis Publishers.

Phipps, R.H., A.K. Jones, A.P. Tingey, and S. Abeyasekera. 2005. Effect of corn silage from an herbicide-tolerant genetically modified variety on milk production and absence of transgenic DNA in milk. *Journal of Dairy Science* 88(8):2870–2878.

Piggott, N.E., and M.C. Marra. 2007. The net gain to cotton farmers of a natural refuge plan for Bollgard II® Cotton. *AgBioForum* 10(1):1–10.

———. 2008. Biotechnology adoption over time in the presence of non-pecuniary characteristics that directly affect utility: A derived demand approach. *AgBioForum* 11(1):58–70.

Pilcher, C.D., and M.E. Rice. 2003. Economic analysis of planting dates to manage European corn borer (Lepidoptera: Crambidae) with Bt corn. *Journal of Economic Entomology* 96(3):941–949.

Price, G.K., W. Lin, J.B. Falck-Zepeda, and J. Fernandez-Cornejo. 2003. Size and distribution of market benefits from adopting biotech crops. November 20. TBN-1906. U.S. Department of Agriculture. Washington, DC.

Qaim, M. 2009. The economics of genetically modified crops. *Annual Review of Resource Economics* 1(1).

Qaim, M., and G. Traxler. 2005. Roundup Ready soybeans in Argentina: Farm level and aggregate welfare effects. *Agricultural Economics* 32(1):73–86.

Raymer, P.L., and T.L. Grey. 2003. Challenges in comparing transgenic and nontransgenic soybean cultivars. *Crop Science* 43(5):1584–1589.

Reddy, K.N., A.M. Rimando, S.O. Duke, and V.K. Nandula. 2008. Aminomethylphosphonic acid accumulation in plant species treated with glyphosate. *Journal of Agricultural and Food Chemistry* 56(6):2125–2130.

Rice, M.E. 2004. Transgenic rootworm corn: Assessing potential agronomic, economic, and environmental benefits. *Crop Management*. Available online at http://www.plantmanagementnetwork.org/pub/php/review/2004/rootworm/. Accessed April 8, 2009.

Rice, M.E., and K. Ostlie. 1997. European corn borer management in field corn: A survey of perceptions and practices in Iowa and Minnesota. *Journal of Production Agriculture* 10(4):628–634.

Richardson, R.J., H.P. Wilson, G.R. Armel, and T.E. Hines. 2004. Mixtures of glyphosate with CGA 362622 for weed control in glyphosate-resistant cotton (*Gossypium hirsutum*). *Weed Technology* 18(1):16–22.

Ronald, P., and B. Fouche. 2006. Genetic engineering and organic production systems. *Agricultural Biotechnology in California*. University of California–Division of Agriculture and Natural Resources. Publication 8188. Available online at http://ucbiotech.org/resources/factsheets/8188.pdf. Accessed March 31, 2010.

Salazar, M.P., J.B. Miller, L. Busch, and M. Mascarenhas. 2006. The indivisibility of science, policy, and ethics: Starlink™ corn and the making of standards. In *Agricultural standards: The shape of the global food and fiber system*. eds. R.J. Bingen and L. Busch, pp. 111–124. Dordrecht, The Netherlands: Springer.

Sanders, L.D. 2000. The economics of conservation and conservation tillage. Paper presented at the International Symposium on Conservation Tillage (Mazatlan, Mexico, January 24–27, 2000). Available online at http://agecon.okstate.edu/isct/labranza/sanders/mazecon00.doc. Accessed August 29, 2009.

Scursoni, J., F. Forcella, J. Gunsolus, M. Owen, R. Oliver, R. Smeda, and R. Vidrine. 2006. Weed diversity and soybean yield with glyphosate management along a north-south transect in the United States. *Weed Science* 54(4):713–719.

Sexton, S., D. Zilberman, D. Rajagopal, and G. Hochman. 2009. The role of biotechnology in a sustainable biofuel future. *AgBioForum* 12(1):130–140.

Sexton, S.S., Z. Lei, and D. Zilberman. 2007. The economics of pesticides and pest control. *International Review of Environmental and Resource Economics* 1(3):271–326.

Shaner, D.L. 2000. The impact of glyphosate-tolerant crops on the use of other herbicides and on resistance management. *Pest Management Science* 56(4):320–326.

Shaw, D.R., and J.C. Arnold. 2002. Weed control from herbicide combinations with glyphosate. *Weed Technology* 16(1):1–6.

Shaw, D.R., and C.S. Bray. 2003. Foreign material and seed moisture in glyphosate-resistant and conventional soybean systems. *Weed Technology* 17(2):389–393.

Siebert, M.W., S. Nolting, B.R. Leonard, L.B. Braxton, J.N. All, J.W. Van Duyn, J.R. Bradley, J. Bacheler, and R.M. Huckaba. 2008. Efficacy of transgenic cotton expressing CrylAc and CrylF insecticidal protein against heliothines (Lepidoptera: Noctuidae). *Journal of Economic Entomology* 101(6):1950–1959.

Sikkema, P.H., C. Shropshire, A.S. Hamill, S.E. Weaver, and P.B. Cavers. 2004. Response of common lambsquarters (*Chenopodium album*) to glyphosate application timing and rate in glyphosate-resistant corn. *Weed Technology* 18(4):908–916.

———. 2005. Response of barnyardgrass (*Echinochloa crus-galli*) to glyphosate application timing and rate in glyphosate-resistant corn (*Zea mays*). *Weed Technology* 19(4):830–837.

Singer, J.W., R.W. Taylor, and W.J. Bamka. 2003. Corn yield response of Bt and near-isolines to plant density. *Crop Management*. Available online at http://ddr.nal.usda.gov/bitstream/10113/11882/11IND43806137.pdf. Accessed March 31, 2010.

Smith, K.R. 2002. Does off-farm work hinder "smart" farming? *Agricultural Outlook* 294: 28–30. Available online at http://www.ers.usda.gov/publications/agoutlook/sep2002/ao2941.pdf. Accessed March 31, 2010.

Smyth, S., G.G. Khachatourians, and P.W.B. Phillips. 2002. Liabilities and economics of transgenic crops. *Nature Biotechnology* 20(6):537–541.

Snow, A.A., D.A. Andow, P. Gepts, E.M. Hallerman, A. Power, J.M. Tiedje, and L.L. Wolfenbarger. 2005. Genetically engineered organisms and the environment: Current status and recommendations. *Ecological Applications* 15(2):377–404.

Stanger, T.F., and J.G. Lauer. 2006. Optimum plant population of Bt and non-Bt corn in Wisconsin. *Agronomy Journal* 98(4):914–921.

Sweet, J., E. Simpson, J. Law, P. Lutman, K. Berry, R. Payne, G. Champion, M. May, K. Walker, P. Wightman, and M. Lainsbury. 2004. Botanical and rotational implications of genetically modified herbicide tolerance in winter oilseed rape and sugar beet (BRIGHT Project). Project report no. 353. Home-Grown Cereals Authority. Cambridge, UK.

Sydorovych, O., and M.C. Marra. 2007. A genetically engineered crop's impact on pesticide use: A revealed-preference index approach. *Journal of Agricultural and Resource Economics* 32(3):476–491.

Tharp, B.E., and J.J. Kells. 1999. Influence of herbicide application rate, timing, and inter-row cultivation on weed control and corn (*Zea mays*) yield in glufosinate-resistant and glyphosate-resistant corn. *Weed Technology* 13(4):807–813.

Thelen, K.D., and D. Penner. 2007. Yield environment affects glyphosate-resistant hybrid response to glyphosate. *Crop Science* 47(5):2098–2107.

Thomas, W.E., I.C. Burke, and J.W. Wilcut. 2004. Weed management in glyphosate-resistant corn with glyphosate and halosulfuron. *Weed Technology* 18(4):1049–1057.

Thomas, W.E., W.J. Everman, J. Allen, J. Collins, and J.W. Wilcut. 2007. Economic assessment of weed management systems in glufosinate-resistant, glyphosate-resistant, imidazolinone-tolerant, and nontransgenic corn. *Weed Technology* 21(1):191–198.

Tingle, C.H., and J.M. Chandler. 2004. The effect of herbicides and crop rotation on weed control in glyphosate-resistant crops. *Weed Technology* 18(4):940–946.

Traore, S.B., R.E. Carlson, C.D. Pilcher, and M.E. Rice. 2000. Bt and non-Bt maize growth and development as affected by temperature and drought stress. *Agronomy Journal* 92(5):1027–1035.

Trigo, E.J., and E.J. Cap. 2003. The impact of the introduction of transgenic crops in Argentinean agriculture. *AgBioForum* 6(3):87–94.

US-EPA (U.S. Environmental Protection Agency). 2001. Biopesticides registration action document—*Bacillus thuringiensis* plant-incorporated protectants. October 16. Office of Pesticide Programs–Biopesticides and Pollution Prevention Division. Washington, DC. Available online at http://www.epa.gov/oppbppd1/biopesticides/pips/bt_brad.htm. Accessed January 19, 2010.

USDA-AMS (U.S. Department of Agriculture–Agricultural Marketing Service). 2000. National Organic Program: Final Rule. *Federal Register* 65(246):80548–80596. Codified at 7 C.F.R. § 205.

USDA-ERS (U.S. Department of Agriculture–Economic Research Service). 2009. Corn: Market outlook, USDA feed grain baseline, 2009–2018. Washington, DC. Available online at http://www.ers.usda.gov/Briefing/corn/2009baseline.htm. Accessed September 2, 2009.

USDA-NASS (U.S. Department of Agriculture–National Agricultural Statistics Service). 2000. Agricultural prices. 1999 summary. July. Pr 1-3 (00) a. Washington, DC. Available online at http://usda.mannlib.cornell.edu/usda/nass/AgriPricSu//2000s/2000/AgriPricSu-07-24-2000.pdf. Accessed April 14, 2009.

———. 2005. Agricultural prices. 2004 summary. July. Pr 1-3 (05) a. Washington, DC. Available online at http://usda.mannlib.cornell.edu/usda/nass/AgriPricSu//2000s/2005/AgriPricSu-08-16-2005.pdf. Accessed April 14, 2009.

———. 2008. Agricultural prices. 2007 summary. July. Pr 1-3 (08) a. Washington, DC. Available online at http://usda.mannlib.cornell.edu/usda/nass/AgriPricSu//2000s/2008/AgriPricSu-07-31-2008_revision.pdf. Accessed April 14, 2009.

———. 2009a. Agricultural prices. 2008 summary. July. Pr 1-3 (09) a. Washington, DC. Available online at http://usda.mannlib.cornell.edu/usda/nass/AgriPricSu//2000s/2009/AgriPricSu-08-05-2009.pdf. Accessed January 31, 2010.

———. 2009b. Data and statistics: Quick stats. Washington, DC. Available online at http://www.nass.usda.gov/Data_and_Statistics/Quick_Stats/index.asp. Accessed June 22, 2009.

USDA-NRCS (U.S. Department of Agriculture–Natural Resources Conservation Service). 2008. Energy Consumption Awareness Tool: Tillage. Washington, DC. Available online at http://ecat.sc.egov.usda.gov/. Accessed May 14, 2009.

Vaughn, T., T. Cavato, G. Brar, T. Coombe, T. DeGooyer, S. Ford, M. Groth, A. Howe, S. Johnson, K. Kolacz, C. Pilcher, J. Purcell, C. Romano, L. English, and J. Pershing. 2005. A method of controlling corn rootworm feeding using a *Bacillus thuringiensis* protein expressed in transgenic maize. *Crop Science* 45(3):931–938.

Venneria, E., S. Fanasca, G. Monastra, E. Finotti, R. Ambra, E. Azzini, A. Durazzo, M.S. Foddai, and G. Maiani. 2008. Assessment of the nutritional values of genetically modified wheat, corn, and tomato crops. *Journal of Agricultural and Food Chemistry* 56(19):9206–9214.

Vermij, P. 2006. Liberty Link rice raises specter of tightened regulations. *Nature Biotechnology* 24(11):1301–1302.

Vogel, G. 2006. Genetically modified crops. Tracing the transatlantic spread of GM rice. *Science* 313(5794):1714.

Wiatrak, P.J., D.L. Wright, J.J. Marois, and D. Wilson. 2005. Influence of planting date on aflatoxin accumulation in Bt, non-Bt, and tropical non-Bt hybrids. *Agronomy Journal* 97(2):440–445.

Wiesbrook, M.L., W.G. Johnson, S.E. Hart, P.R. Bradley, and L.M. Wax. 2001. Comparison of weed management systems in narrow-row, glyphosate- and glufosinate-resistant soybean (*Glycine max*). *Weed Technology* 15(1):122–128.

Williams, W.P., G.L. Windham, P.M. Buckley, and J.M. Perkins. 2005. Southwestern corn borer damage and aflatoxin accumulation in conventional and transgenic corn hybrids. *Field Crops Research* 91(2-3):329–336.

Wilson, T.A., M.E. Rice, J.J. Tollefson, and C.D. Pilcher. 2005. Transgenic corn for control of the European corn borer and corn rootworms: A survey of midwestern farmers' practices and perceptions. *Journal of Economic Entomology* 98(2):237–247.

Wu, F. 2006. Mycotoxin reduction in Bt corn: Potential economic, health, and regulatory impacts. *Transgenic Research* 15(3):277–289.

Wu, F., J.D. Miller, and E.A. Casman. 2005. Bt corn and mycotoxin reduction: An economic perspective. In *Aflatoxin and food safety*. ed. H.K. Abbas, pp. 459–482. Boca Raton, FL: CRC Press.

4

Farm-System Dynamics and Social Impacts of Genetic Engineering

The dissemination of genetically-engineered (GE) crops, like the adoption process associated with other farm-level technologies, is a dynamic process that both affects and is affected by the social networks that farmers have with each other, with other actors in the commodity chain, and with the broader community in which farm households reside. As noted in Chapter 1, farmer decisions to adopt a technology are influenced not only by human-capital factors, such as the educational level of the adopter, but by social-capital factors, such as access to information provided by other farmers through social networks (Kaup, 2008). That necessarily implies that farmers receive information from others—for example, on the risks and benefits of a particular technology—and that they share their own knowledge and experience through the same networks. Such findings confirm the relevance of social factors in influencing how genetic-engineering technology is adopted, what the impacts of its adoption are, and the significance of farmers' active participation in both formal and informal social networks with other actors in commodity chains and communities.

However, little research has been conducted on the social impacts of the adoption of genetic-engineering technology by farmers, even though there is substantial evidence that technological developments in agriculture affect social structures and relationships (Van Es et al., 1988; Buttel et al., 1990). Because further innovations through genetic engineering are anticipated, such research is needed to inform seed developers, policy makers, and farmers about potential favorable benefits for adopters and nonadopters and unwanted or potentially unforeseen social effects (Guehlstorf, 2008). With such information, the likelihood of maximizing social benefits while minimizing socials costs is increased. To demonstrate the necessity for increasing commitments to the conducts of research on the social effects of GE-crop adoption, this chapter synthesizes what is known in the scientific literature about the social impacts of farm-technology adoption and the interactions between farmers' social networks. The chapter also identifies future research needs.

SOCIAL IMPACTS OF ON-FARM TECHNOLOGY ADOPTION

The earliest academic research in the United States on the social impacts of technology adoption at the farm and community levels was focused on mechanical technologies. More than a century ago, the use of machines in U.S. agriculture not only displaced labor but widened socioeconomic discrepancies between skilled and unskilled laborers (Quaintance, 1984). Academic interest in the socioeconomic consequences of agricultural mechanization was particularly strong in the 1930s and 1940s in the southern United States (Buttel et al., 1990) and again in the 1970s throughout the country. Berardi (1981) summarized the findings of the literature and found that mechanization was associated with decreases in the agricultural labor force, particularly those among the least educated and least skilled workers and in minority groups; with better working conditions and less "drudgery" for the remaining work force; with a decrease in farm numbers and an increase in farm size; with increased capital costs for agricultural producers; and with a decline in the socioeconomic viability of agriculture-dependent rural communities. Data also suggested that the technological development of U.S. agriculture had contributed to declines in farm labor, in community dependence on agriculture, and in rural community viability although other on-farm and off-farm factors also contributed to these changes (Van Es et al., 1988).

In the 1980s, social scientists broadened their research on the impacts of technology adoption on farms and farm communities to include studies of the potential and actual impacts of biological (pregenetic engineering) technologies in agriculture. Many observers assumed that, unlike the earlier wave of mechanical agricultural technologies, genetic-engineering technology would not be biased towards large-scale farming operations. Such an assumption was supported by analyses of the production capabilities of agricultural biotechnology. For example, it was noted that no interaction effect was observed between genetic predisposition to produce milk and the use of the growth hormone bovine somatotropin (BST) to increase milk production in dairy cows (Nytes et al., 1990). However, other studies that directly examined farm-level social change revealed that, despite the presumption of scale-neutrality, it was difficult to isolate the impacts of biological innovations from those of other technological innovations in agriculture because biological innovations were often developed and disseminated in conjunction with other technologies that may not have been scale-neutral (Kloppenburg, 1984).

Additional research conducted on the social impacts of biotechnology in animal agriculture, specifically on the use of BST, noted that rates of adoption of BST were moderate and that, although adoption did not require large herds, scale effects were observed because BST use was more effective in high-producing cows, which were more likely to be found in large herds with complementary feeding technologies (Barham et al., 2004). Beck and Gong (1994) also observed the existence of a scale effect with adoption of BST, with adopters more likely to have larger herds, as well as being younger and having more formal education. Additionally, it was suggested that the quality of farm management had an impact on the benefits accruing to the adoption of BST (Bauman, 1992). The use of BST also was thought to lead to lower prices and thus to result in increased economic pressure on smaller producers (Marion and Wills, 1990). In other words, the body of research on the socioeconomic consequences of the use of biotechnologies, including Green Revolution technologies, indicated that "scale neutrality is not inevitable, but a possibility that depends on institutional context" (DuPuis and Geisler, 1988: 410). To put it another way, the social context of the adoption process and the impacts on that context are interconnected, from which it follows that the social impacts of genetic-engineering technology on farms and communities differ among cultures, commodities, and historical periods.

Thus, though seed varieties are generally conceptualized as being scale-neutral, the adoption of any technology may be biased toward large firms that can spread the fixed costs of learning over greater quantities of production (Caswell et al., 1994). In developing countries, the economics of genetic-engineering technology do not appear to vary with farm size (Thirtle et al., 2003). However, scale may affect accessibility to technology. Small farmers have less influence in input supply and marketing chains with which to secure access to desired technologies. Thus, there can be a scale bias in the development and dissemination processes associated with herbicide-resistance technology that puts small farmers at a disadvantage. In contrast, as noted in Chapter 3, insect-resistance technology can replace insecticide applications that require fixed capital investments, such as for tractors and sprayers. In this regard genetic-engineering technology has the potential to favor small farmers, who would benefit more from a technology that required less fixed capital investment. The scale effects of transgenic varieties may also depend on the pricing (such as quantity discounts) set by seed companies, which typically assess a technology-user fee.[1]

An early empirical study was carried out by Fernandez-Cornejo et al. (2001) using 1998 U.S. farm data. They found that, as expected, the adoption of HR soybean was invariant to size, but adoption of HR corn was positively related to size. They explained this disparity as due to the different adoption rates: 34 percent of the farms had adopted HR soybean at the time, implying that adoption of HR soybean had progressed passed innovator and early-adopter stages into the realm where adopting farmers are much like the majority of farmers. On the other hand, adoption of HR corn was quite low at the time (5 percent of farms), implying that adoption was largely confined to innovators and other early adopters who in general tend to control substantial resources and who are willing to take the risks associated with trying new ideas. Thus, they claimed that the impact of farm size on adoption is highest at the very early stages of the diffusion of an innovation (HR corn), and becomes less important as diffusion increases. This result confirms Rogers's (2003) observations that adoption is more responsive to farm size at the innovator stage, and the effect of farm size in adoption generally diminishes as diffusion progresses. Early adopters, by virtue of early adoption, also are able to capture a greater percentage of the economic benefits of the technology adoption process.

Clearly, one cannot extrapolate the social impacts of the adoption of GE crops based solely on an assumption that the productive capabilities of genetic-engineering technology, when isolated from the interaction with other factors, should be scale-neutral. In other words, previous research on the social impacts of agricultural technologies suggests the possibility that the early dissemination of genetic-engineering technology would be associated with farm size, and that the use of GE crops could have differential impacts across farm types, farm size, and region, despite the fact that GE crops are presumed to be scale-neutral.

In an article that attempted to predict some of the environmental, economic, and social effects of genetic engineering of crops, it was argued that the use of GE crops was "clearly capable of causing major ecological, economic, and social changes" (Pimentel et al., 1989: 611). Nonetheless, over the last decade, there has been virtually no empirical research conducted on the social impacts of the use of GE crops on farms and rural communities. The lack of research may have to do in part with the scarcity of funds available for such research as well as a relative lack of interest in social issues on the part of environmental groups (Chen and

---

[1]Examples of empirical studies on the effect of farm size on GE-crop adoption are given in "An Early Portrait of Farmers Who Adopt Genetically Engineered Crops" in Chapter 1.

Buttel, 2000), and other groups and organizations that might be expected to support such research. Nonetheless, the results of research referred to above on the social repercussions of agricultural technologies, including non–genetic-engineering biotechnology in crops and biotechnology in animal agriculture, would suggest that there are impacts, that these impacts could be favorable or adverse, and that adverse impacts could be alleviated through the adoption of appropriate policies. For example, based on earlier research on the introduction of new technologies in agriculture, it might be hypothesized that certain categories of farmers (those with less access to credit, those with fewer social connections to university and private sector researchers, etc.) might be less able to access or benefit from existing GE crops. There is also the possibility that the types of genetic advances being marketed do not meet the needs of certain classes of farmers, and that the full spectrum of the potential of genetic-engineering technology is not being achieved. Furthermore, the possibility exists that communities where farmers play an important social, political, and economic role could be impacted as well. However, for the purpose of this report, no conclusion on the social impacts of the adoption of GE crops can be drawn on the basis of empirical evidence. Research on such impacts clearly should be accorded a high priority as genetic-engineering technology evolves. Without such research, the potential for genetic-engineering technology to contribute to the sustainable development of U.S. agriculture and rural communities cannot be adequately assessed. Thus, we recommend that such research be sponsored and pursued actively and immediately.

SOCIAL NETWORKS AND ADOPTION DECISIONS

The adoption of genetic-engineering technology and its performance on the farm are functions of the knowledge of agricultural decision makers, who include farmers, input suppliers, commodity traders, farm-management consultants, and extension agents. In making technology-adoption decisions, farmers rely principally on information about the relative performance of competing technologies and on information about best practices for optimizing yields and controlling costs, given the technologies that they use. The performances of firms and technology, therefore, depend upon the information used by various commodity-system actors. Just et al. (2002) have shown that the internal competences of decision makers affect the degree to which they rely on different types and sources of information.

Farmers rely on a variety of intermediaries—such as extension agents, commodity groups, commercial vendors, agricultural media, and other farmers—for information. For example, farmers often turn to commodity associations for information about regulations and regulatory changes. Many of the intermediaries that farmers communicate with use public information, especially data and research results provided by the U.S. Department of Agriculture (USDA) Economic Research Service and the National Agricultural Statistics Service and by state extension services, particularly for information about the economic outlook of agriculture and specific industries. Intermediaries use formal channels of information more than farmers (Just et al., 2002) and then make that information available to farmers.

Farmers obtain about half their information from informal sources (i.e., sources whose professional duties do not include provision of information) (Just et al., 2002), including people in the end-users' civic, community, professional, and commercial networks, like neighbors, colleagues, customers, and suppliers. Farmers' reliance on informal sources may reflect low availability of or access to information from formal channels, issues of affordability of private information, and credibility (Just et al., 2002).

Those findings suggest that farmers' attitudes toward GE crops are likely to be affected by a number of information providers. USDA's Cooperative Extension Service, commodity groups, and agricultural media are particularly influential in informing farmers' views on the technical aspects of genetic-engineering technology, its economic implications, and its prospects. Although the influence of those sources has not been widely appreciated, they have played a key role in the adoption of the technology. As Wolf et al. (2001) and Just et al. (2002) demonstrated, informal sources of information are just as likely as formal sources to accelerate or to slow the rate of GE-crop adoption. It would also be reasonable to hypothesize that patterns of information use would be linked to the ability of farmers to use the technology effectively and maximize its potential.

INTERACTION OF THE STRUCTURE OF THE SEED INDUSTRY AND FARMER DECISIONS

The U.S. seed industry has experienced extensive structural change in the last few decades. The changes have affected decisions at the farm level by shaping the choices available to corn, soybean, and cotton farmers.

As Fernandez-Cornejo and Just (2007) have summarized, plant-breeding research until the 1930s was conducted primarily by the public sector (for example, USDA and state agricultural experiment stations), and most commercial seed suppliers were small, family-owned businesses that multiplied seed varieties that had been developed in the public domain. Seeds embody the scientific knowledge needed to produce a new plant variety with desirable attributes—such as higher yield, disease or pesticide resistance, or improved quality—so seed innovators face both the risk of imitation by competing seed firms and the risk of seed reproduction by farmers themselves (Fernandez-Cornejo, 2004). The development of hybrid corn in the first half of the 20th century provided breeders with greater protection of intellectual-property rights (IPR) because seeds saved postharvest produced substantially smaller yields than the hybrid plants from which they were gathered. With that incentive, the number of private firms engaged in corn breeding grew rapidly.

The proliferation of firms was followed by consolidation in part because U.S. law evolved to provide incentives to innovators for research and development by giving them exclusive control of their innovations through patent laws and other forms of enforceable legal protection (Fernandez-Cornejo, 2004). Two principal forms of legal protection for seed innovators are plant variety protection (PVP) certificates issued by the Plant Variety Protection Office of USDA and patents issued by the Patent and Trademark Office (PTO) of the U.S. Department of Commerce. Both grant private crop breeders exclusive rights to multiply and market their newly developed varieties. Patents provide more control because PVP certificates have a research exemption that allows others to borrow a new variety for research purposes (Fernandez-Cornejo and Schimmelpfennig, 2004). IPRs for seed innovators were strengthened by the U.S. Supreme Court's 1980 *Diamond* v *Chakrabarty* decision, which extended patent rights to GE microorganisms, important tools and products of biotechnology. A series of rulings by PTO's Board of Appeals and Interferences widened the scope of patent protection for GE organisms by including plants and nonhuman animals. Those rulings extended IPRs to a wide array of new biotechnology products in the form of utility patents (also referred to as patents for invention). Products protected under the rulings include seeds, plants, plant parts, genes, traits, and biotechnology processes (Fuglie et al., 1996; Fernandez-Cornejo, 2004).

Enhanced IPR protection has brought rapid increases in private research and development (R&D), and indirectly assisted technology developers in setting prices above marginal costs (Goldsmith, 2001). Private spending on R&D in crop varieties increased by a factor of 14 in real terms from 1960 to 1996 (Fernandez-Cornejo, 2004), whereas public (federal and state) spending changed little (Figure 4-1; Fernandez-Cornejo and Schimmelpfennig, 2004). At the same time, IPR protection may have spurred market concentration in the seed industry. The potential profits of seed firms made possible through IPR protection may strengthen the incentive to invest and thus provide greater opportunities to large firms (Lesser, 1998). Many seed firms have been acquired by corporations that have the resources needed to achieve large economies of scale in R&D (Fernandez-Cornejo, 2004). For example, Lesser (1998) stated that more

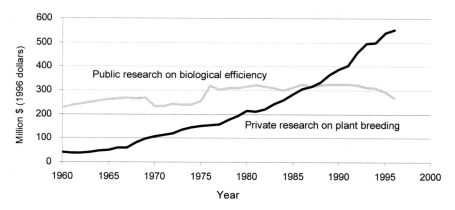

FIGURE 4-1 Public and private research expenditures on plant breeding. Biological efficiency includes breeding and selection of improved plant varieties. SOURCE: Fernandez-Cornejo, 2004.

than 50 seed firms were acquired by pharmaceutical, petrochemical, and food firms after the passage of the Plant Variety Protection Act.[2] In contrast, Lesser (1998) also noted that weakness of IPR protection may lead to mergers and acquisitions. In any case, by 1997, the share of U.S. seed sales controlled by the four largest firms reached 69 percent for corn (up from 60 percent 1973), 47 percent for soybean (up from 7 percent in 1980), and 92 percent for cotton (up from 74 percent in 1970) (Table 4-1; Fernandez-Cornejo, 2004). Though it is difficult to obtain recent detailed published market share information, it appears from company reports and other sources that the trend of increased concentration in the structure of the seed industry continued in recent years.[3] Farm survey data for corn and soybean indicated that by 2007 the share of the four largest firms reached 72 percent for corn and 55 percent for soybean (Figure 4-2; Shi and Chavas, 2009).

Concentration of R&D output can also be used to measure the concentration in innovation activity in the seed industry (Fulton and Giannakas, 2001). In genetic-engineering technology, a measure of R&D output is the

---

[2]The Plant Variety Protection Act (PVPA) of 1970 granted plant breeders a certificate of protection that gave them exclusive rights to market a new plant variety for 18 years from the date of issuance. Amendment of the PVPA in 1994 brought it into conformity with international standards. Protection provided by certificates of protection was extended from 18 to 20 years for most crops (Fernandez-Cornejo, 2004).

[3]In the case of corn, Pioneer has lost its dominant position in the corn seed market from about 40 percent to 30 percent while Monsanto's share of the corn seed market increased to about 30 percent as a result of the Landec acquisition (Leonard, 2006).

TABLE 4-1 Estimated Seed Sales and Shares of Major Field Crops, United States, 1997

Company	Total ($ billions, current)	Corn Market Share	Soybean Market Share	Cotton Market Share
		Percentage of Acres		
Pioneer Hi-Bred International	1.18	42.0	19.0	—
Monsanto/Stoneville	0.54	14.0	19.0	11.0
Novartis/Syngenta	0.26	9.0	5.0	—
Delta & Pine Land	0.08	—	—	73.0
Dow Agrosciences/Mycogen	0.14	4.0	4.0	—
Golden Harvest	0.09	4.0	—	—
AgrEvo/Cargill	0.09	4.0	—	—
Others	1.12	23.0	53.0	16.0
Total	3.50	100.0	100.0	100.0

SOURCES: Hayenga, 1998; Fernandez-Cornejo, 2004.

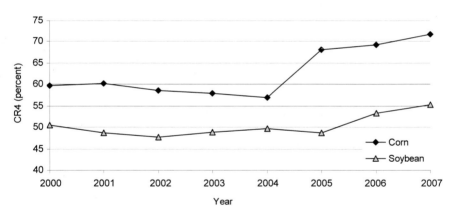

FIGURE 4-2 Share of planted acres of corn and soybean seeds by largest four firms (CR4).
SOURCE: Stiegert et al., 2009.

number of GE cultivars approved by USDA for release into the environment for field testing. In particular, Fernandez-Cornejo (2004) adapted the four-firm concentration-ratio measure, commonly used to quantify industry concentration in terms of sales, to examine R&D concentration on the basis of regulatory approvals of GE crop varieties. Table 4-2 shows the percentage of field releases obtained by the leading four firms

TABLE 4-2 Four-Firm Concentration Ratio in Field-Release
Approvals from USDA Animal and Plant Health Inspection Service,
by Crop, 1990–2000

	1990	1991	1992	1993	1994	1995	1996	1997	1998	1999	2000
Corn	67	67	65	82	82	67	60	73	73	80	79
Soybeans	100	100	94	68	72	94	82	82	71	87	85
Cotton	100	100	100	89	79	85	91	64	98	98	96

SOURCE: Fernandez-Cornejo, 2004.

in 1990–2000. The top four firms controlled well over 50 percent of the
approvals; this suggests consolidation in R&D and potential barriers to
entry for competitors. As Fulton and Giannakas (2001) noted, expenditures on R&D and expenditures made to obtain regulatory approvals are
sunk costs—costs that cannot be recouped. If such sunk costs are present,
markets are not contestable, so there are potential barriers to entry.[4]

As Fernandez-Cornejo (2004) observed, on the basis of the four-firm
concentration ratio of approvals, the extent of corn-seed R&D concentration has been relatively constant at around 72 percent, which is fairly
consistent with the four-firm concentration ratio in corn in terms of sales.
Cotton-seed R&D is the most centralized, and this is also consistent with
market-concentration measures.

Patent ownership shows a pattern of concentration similar to that evi-
dent in other R&D measures (Fernandez-Cornejo, 2004). Most of the biotechnology patents awarded to private firms are held by a small number
of large companies. As of 1996–1997, Pioneer (soon after DuPont/Pioneer)
held the largest number of patents for corn and soybean, followed by
Monsanto (Brennan et al., 1999). The leading firms in the sector have
received IPR protection not only by virtue of their respective R&D investments but through mergers and acquisitions. For example, Pioneer was
one of the first four companies active in the emerging corn-seed market in
the early 1930s. As shown in Figure 4-3, Pioneer (Pioneer Hi-Bred International, Inc.) made a series of acquisitions in 1973–1980 that strengthened
its overall position in the seed market. The chemical firm DuPont bought
20 percent of Pioneer in August 1997 and bought the remaining 80 percent in 1999 for $7.7 billion. As a DuPont company, Pioneer continues
to operate under the Pioneer name and remains headquartered in Iowa
(Fernandez-Cornejo, 2004).

Although the increase in seed-industry concentration has raised concerns about its potential impact on market power, and ultimately on

---

[4]A contestable market behaves in a competitive manner despite having few companies
because of the threat of new entrants.

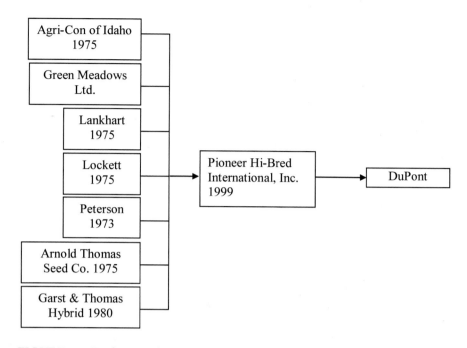

FIGURE 4-3 Evolution of Pioneer Hi-Bred International, Inc. / E.I. du Pont de Nemours and Company.
SOURCE: Fernandez-Cornejo, 2004.

the sustainability of farms, empirical results for U.S. cotton and corn seed industries over the period covering 1970–1998 (which includes only 2 years of GE-crop adoption) suggest that increased concentration during that time period resulted in a cost-reducing effect that prevailed over the effect of enhanced market power (Fernandez-Cornejo, 2004). Goldsmith (2001) argued that even though GE-seed prices were above the competitive price, the actions of biotechnology supply firms apparently were not adversely affecting the welfare of U.S. farmers.

However, concerns have been raised that, in time, such market power could lead to decreased variability in the types of seeds being produced for the market, as well as increased prices, which could limit the ability of farmers to purchase those seeds most suited for local environmental conditions. In addition, it is conceivable that the continued market power of biotechnology supply firms could lead to increased input costs for farmers, which in turn could have an unfavorable effect on the socioeconomic sustainability of farms. A recent study by Shi and Chavas (2009) has found that vertical integration (ownership of control of different stages of production) in the U.S. soybean seed industry had a substantial effect on soybean prices. Shi et al. (2008) analyzed the pricing of corn seed with stacked traits for the years 2000–2007 and found "significant departures from component pricing (where seeds are priced as the sum of their component values). The evidence supports sub-additive pricing. It shows that the marginal contribution of each component to the seed price declines with the number of components." The authors also have indicated that "such a finding is consistent with the presence of economies of scope in seed production. Indeed, synergies in R&D investment (treated as fixed cost) across seed types can contribute to reducing total cost."

In response to these concerns and others, USDA and the U.S. Department of Justice launched a series of workshops in 2010 to examine competition and regulatory issues in the agriculture industry (USDA, 2010). This is a first step towards updating and continuing research on how market structure in the seed industry may be impacting seed prices and availability to variability in genetic resources. In addition, studies of how seed-industry concentration, as well as the practice of cross-licensing, could interact with farmers' planting options and decisions, overall yield benefits, crop genetic diversity, and economic returns would be very valuable.

Although the private sector owns the majority of agricultural-biotechnology patents, the public sector still owns a substantial share. In a study of assignment of U.S. plant-biotechnology patents granted from 1982 to 2001, Graff et al. (2003) found that 41 percent of the patents were owned by large biotechnology companies, 35 percent by startups, and 24 percent by the public sector. The public-sector ownership is weaker in some categories, such as Bt and other insect-resistant traits (10 percent) and plant enzymes (8 percent), but stronger in other categories (42 percent in flowering and 56 percent in pathogen resistance). The capacity of the public sector to obtain freedom to operate in transgenic crops is often also constrained by the fragmentation of technology ownership among numerous institutions. Improved access to information about IPRs and reduced transaction costs to obtain rights to use patents could increase the public sector's contribution to the development of transgenic varieties. Concerns have also been raised that technology use and stewardship agreements prevent scientists in the public sector from conducting independent assessments of GE varieties marketed by the private sector. In February 2009, in response to a notice in the *Federal Register* on a meeting of the Federal Insecticide, Fungicide, and Rodenticide Act Scientific Advisory Panel, 26 entomologists submitted a general comment that, because of those restrictions, the data that the Environmental Protection Agency received regarding GE crops were inherently limited (US-EPA, 2008). If such restrictions exist, farmer welfare could be adversely affected by the lack of complete information regarding GE traits or crops. However, the degree to which technology use agreements may hamper public research is unclear and strongly disputed by private-sector seed companies (Monsanto, 2010). This issue merits careful investigation by neutral researchers to understand what, if any, effects such agreements have on public research.

SOCIAL AND INFORMATION NETWORKS BETWEEN FARMERS AND INDUSTRY

Agriculture is unique among American industries in that federal law allows farmers to cooperate in some collective activities while competing in the output market. That has enabled farmers to act collectively as a counterweight to the large firms on which they rely to sell their output (Cochrane, 1993). Farmers have developed cooperatives to coordinate research and marketing efforts to benefit from economies of scale in these activities (Sexton, 1986). That collective-action capacity is important for adoption of genetic-engineering technology in that it gives farmers the ability to influence crop traits that are introduced.

Farmers may attempt to block seed technologies if they anticipate that those new technologies will not enhance, or perhaps even endanger, farm profit. Innovation in any crop can increase farm revenue if the elasticity of demand for the crop is sufficiently high that increases in output compensate for decreases in prices. But if the elasticity of demand is low, increased output may lead to a fall in revenue and a decrease in profit unless the technology delivers cost savings. In the latter case, farmers may attempt to block the introduction of a technology. Farmers may also block a technology if its introduction would result in lower prices in national or international markets.

A large body of literature on the political economy of research argues that farmers use political pressure to shape public research funding. Ruttan (1982) argued that farmer pressure may have led to underinvestment in public research in the United States, and de Gorter and Zilberman (1990) linked overall spending on research to the political power of such groups as consumers and producers. Graff and Zilberman (2007) argued that farmers' interests partly motivated Europe's effective ban on GE crops, which in turn affected the access of U.S. farmers who were growing GE crops to European markets.

Another example of farmers' collective action is how farmers' concerns influenced Monsanto's decision to halt its efforts to introduce and market herbicide-resistant (HR) wheat. Some farmer coalitions in the United States and in Canada played a role in that decision because they feared losing access to European and Asian markets that would not accept

HR wheat. It was thought that the introduction of that GE crop might have closed markets to U.S. and Canadian farmers who planted non-GE wheat because of the difficulty in segregating wheat on fields and in grain elevators and trucks (e.g., Anonymous, 2001; Pollack, 2004). Many international buyers, including millers and bakers in important Asian and European markets, had made it clear that they wanted to purchase identity preserved (IP), non-GE commodities (Vandenberg et al., 2000). In the case of white wheat, Japan and South Korea, two countries that have GE-food labeling laws, were importing more than one-fourth of U.S. white wheat exports (Squires, 2004). Japanese millers and bakers were well aware that a large percentage of Japanese consumers were expressing negative attitudes toward the use of genetic-engineering technology in foods and believed that their government's regulation of GE food was too lax (Toyama et al., 2001). Indeed, Japan was the United States's largest market for non-GE corn and soybean: Nearly 90 percent of the IP, non-GE corn and soybean produced in the United States was being exported to Japan (Wilson et al., 2003). Faced with such demands from the marketplace and with farmer concern about how those demands might influence sales in those markets, Monsanto decided to defer the introduction of HR wheat in 2004.

However, farm organizations are not monolithic; indeed, the issue of HR wheat divided the farming community (Graham and Martin, 2004). Even after Monsanto chose to suspend its HR-wheat program, wheat producers and support groups, such as the National Association of Wheat Growers, exhibited a strong interest in glyphosate-resistant wheat (Jussaume et al., 2004). Monsanto's HR trait had been inserted into only one variety of wheat grown in a subset of U.S. and Canadian regions. It was opposed primarily by farmers who did not plant the potential HR variety and by some farmers who could have planted it but were afraid of losing access to European and Asian markets. Furthermore, during its deliberations, the present committee heard that North Dakota wheat farmers who were against the introduction of HR wheat because of concern about losing market access in Europe may now believe that they are disadvantaged by not having HR varieties and would like to see government and industry action that would lead to its development (Wilson, personal communication). Pest pressure is heterogeneous among growing regions, so support for or opposition to damage control through GE traits will vary according to farmers' abilities to benefit from them. Thus, the potential for collective action to restrain the power of the seed industry is a function of farmers' common interests, which are often variable.

Farmer cooperatives may also undertake efforts to bring about the introduction of seed traits that the private sector, for economic or other reasons, is not motivated to introduce. They may pool resources to undertake their own research or to defer the regulatory and development costs that private firms face when they introduce new seed technologies. Farmers have worked with universities to introduce new technologies (Bradford et al., 2006), and similar collaborations could be effective in the development of genetic-engineering technology. In California, grape growers have suffered considerable losses from Pierce's Disease, so they have contributed funds to the Public Intellectual Property Resource for Agriculture to support research for a genetic-engineering solution to the problem (PIPRA, 2006). More generally, pooled funds from farmers can lead to the introduction of desired traits for specialty crops when private seed companies lack the incentives to develop the traits alone (for more discussion of seed-access issues, see Chapter 3). As the price of wheat goes up, for instance, farmers in some regions may approach companies or universities to develop seed varieties to address specific constraints on productivity (Wilson, personal communication). They may also work with initiatives like the Specialty Crop Regulatory Assistance program, a fledgling collaboration of the federal government, scientists in public universities and the private sector, and farmers designed to assist technology developers in negotiating GE specialty crops through the requirements and expense of the regulatory process.

Despite the ability of farmers to organize collectively to counterbalance seed companies and processors, some farmers are concerned about the evolution of seed-technology innovation from a public good to a private good that is controlled by firms that have market power derived from patents on specific products. Innovation in most agricultural inputs is embodied in the technology, as in tractors and fertilizers. The technology is developed and sold by the private sector. But because it had historically been difficult to capture benefits from research efforts in seed technology before the advent of hybrid corn seed, the private sector underprovided seed-technology innovation, and the public sector took the lead in providing improved seed varieties in many crops (especially wheat, soybean, cotton, barley, and oat). Consequently, farmers who grew those crops—particularly wheat, barley, and oat—became accustomed to free or low-cost access to seed, and some farmers may consider open access to seeds a right.

In the case of GE seed, however, companies can make use of patent protection to enforce contracts that disallow reuse of the seed grown in farmers' fields. Farmers must instead purchase seed from firms to reward their research investment and effort. The patents enable the private sector to set prices for the protected technology and to restrict the flow of knowledge. The private sector had already begun to invest more in seed technology for major crops such as corn in the 1930s and soybean in the 1970s. Because of the difficulty and expense of removing lint from the seed, cotton farmers had traditionally purchased their seed from ginners and seed distributors. Consequently, the introduction of GE seeds by the private sector and the patented nature of the technology in the case of commercially available corn, soybean, and cotton may not have appeared to be strikingly different from the established relationship between seed companies and farmers of those commodities.

However, the developmental trajectory of GE-seed technology is leading to concern that access to seeds without GE traits or to seeds that have only the specific GE traits of particular interest to farmers may become increasingly limited. Additional concerns are being raised about the lack of farmer input and knowledge regarding which seed traits might be developed. The push to develop seed varieties with a series of stacked traits, some of which may not be of use to some farmers with respect to short-term productivity (leaving aside the issue of improved resistance management discussed in Chapter 2), raises the issue of access to seeds that have equivalent yield potential but only the desired GE traits or no GE traits at all. Although the committee was not able to find published research that documents the degree of U.S. farmers' access to and the quality of non-GE seed, testimony provided to the committee suggested that access to non-GE or nonstacked seed could become limited for some farmers and that available non-GE or nonstacked seed may not have the same yield characteristics as GE cultivars (Hill, personal communication). Research is needed to investigate the extent to which U.S. farmers are having difficulty purchasing high-yielding, non-GE seed. Public-sector institutions could address this concern by improving the design of licensing contracts with seed companies so that property rights of privately developed traits or cultivars will revert to university research programs if private companies do not use the technologies.

Boehlje (1999) has suggested that U.S. agriculture is going through a structural change in which activities that will enhance product differentiation and added value to farming are being emphasized. As part of that evolution, many agricultural sectors (poultry, swine, and some fruits and vegetables) have come to be dominated by contracting arrangements between major agribusiness companies and farmers or by large vertically integrated agribusiness firms. Those large companies have the resources and scale to finance research in the development of GE traits. The emergence of alliances between biotechnology companies and large agribusiness firms, and even large farmers' cooperatives to produce proprietary GE varieties, appears possible, but future research is needed to determine whether such relationships can lead to the development of differentiated products (Boehlje, 1999)—including those with traits that enhance direct value to consumers, such as improving health or convenience, or that respond to the environmental and management needs of specific groups of farmers—and whether such relationships will limit farmers' access to the types of GE traits they value.

INTERACTIONS OF LEGAL AND SOCIAL ISSUES SURROUNDING GENETIC ENGINEERING

Legal issues constitute an important sociopolitical dimension that influences the adoption of genetic-engineering technology and its impacts on farmers and communities. The legal issues are complex, and a complete treatment of them is beyond the expertise of any of the authors of this report. We briefly touch here on the issues of seed saving, gene flow, and organic standards.

Seed Biotechnology

Courts in the United States and Canada have consistently upheld the rights of companies that sell patented seeds and genes through technology-use agreements to prohibit seed-saving practices that involve seed sold through those contracts (Kershen, 2004; Anonymous, 2008). Although that property right has been established, some continue to express concern about the ethical issues surrounding the patenting of life forms and over the effects of technology-use agreements on seed-saving practices. Research on whether those concerns are warranted and what the impacts are on farm sustainability are needed. Concerns are also being raised about the lack of farmer involvement in GE-trait development for traits that could address production problems identified by farmers and over the implications of current patenting procedures on power relationships between biotechnology firms and farmers (Phillipson, 2001). However, the social and economic effects of the exercise of such property rights, especially actual or potential litigation on both adopters and nonadopters of GE crops, have not been thoroughly investigated by social scientists. The lack of academic analyses of those issues may be due in part to the fact that companies, in any sector, that use the courts to enforce their property rights view legal actions and any out-of-court settlements as proprietary information. One interesting response by those who are concerned about the possible effects of the private control of genetic resources has been the open-source breeding movement.[5]

---

[5]This movement, which has been inspired in part by open-source project movements in computer software and elsewhere, is in essence an attempt to develop publicly available genetic resources. As in the case of "shareware," researchers working on open-source biotechnology can access and improve on publicly available genetic resources and technologies but must agree to make the improved materials available for others to use (Delmer, 2005; Lerner and Tirole, 2005).

Gene Flow

A second set of legal issues related to genetic-engineering technology has to do with gene flow, particularly from fields of GE crops to those managed by people not using GE crops (for more on the potential for gene flow between GE and non-GE crops and on the challenges of coexistence of GE and non-GE crops, see Chapter 2). As in cases that involve restrictions on farmers against seed saving, these issues can be viewed as property-rights issues. Does gene flow impinge on the rights of producers and consumers who wish to grow and eat foods that do not include GE material (Conner, 2003)? That is of particular concern to some farmers who wish to produce organic or non-GE crops. Even though organic certification by the U.S. government is determined by the process used to grow the crop, some farmers are concerned that their products may not be accepted by markets in other countries or by food distributors and consumers who establish their own standards, irrespective of process.

Several lawsuits have been filed by farmers against agricultural-biotechnology companies in part because of damage alleged to have occurred as a result of drift of genetic material to the fields of farmers who do not wish to grow crops with GE traits (Kershen, 2004). Consumer groups have also brought legal action against the federal government for approving the commercialization of GE crops that have the potential to cross with non-GE crops in the same vicinity. As was discussed in previous chapters, GE alfalfa was pulled from the market after a U.S. federal judge sided with arguments brought forward by numerous plaintiffs and found that USDA should have prepared an environmental impact statement before it deregulated the crop, which facilitates commercialization (*Geertson Farms v. Johanns*, 2009). In another case filed by the Center for Food Safety and other plaintiffs, a federal judge decided in September 2009 that similar steps should have been taken before GE sugar beet was commercialized (Pollack, 2009).

Issues raised by the possibility of gene flow are not only legal in nature. As noted in Chapters 2 and 3, the adventitious presence of GE material in non-GE crops raises complex environmental and economic challenges. Similarly, social problems could arise as a consequence of gene flow, particularly if GE and non-GE producers of the same commodity live in the same community. Gene-flow disputes could move beyond the merely legal and affect the overall functioning of communities where such disputes exist. This might include conflicts between farmers as well as stress related to the economic and social costs associated with lawsuits and the potential threat of lawsuits. Studies of the social effects of such disputes are needed to gauge the full impact on community well-being. The ability of GE production and non-GE production to coexist in society may depend on the health of communities. Proposals for establishing "landscape clubs" (Furtan et al., 2007) and voluntary "GMO-free zones"[6] (Jank et al., 2006) clearly depend on the existence of high levels of community cooperation, which could be undermined by disputes related to gene flow.

Organic Laws and Resistance to Genetic Engineering

One of the intriguing public debates that has emerged around genetic engineering in agriculture has been that regarding whether GE crops should be allowable in legal standards for organic agriculture. As discussed in Chapter 1, many organic growers have vehemently resisted the notion that GE crops should be allowable in organic agricultural production systems. However, scientific arguments can be made for the use of genetic-engineering technology for making organic agricultural production more sustainable. Ronald and Adamchak (2008) note that what is or is not an appropriate use of genetic-engineering technology for "organic" producers is problematic given that genetic-engineering techniques can be used to transfer genes within plant species as easily as between them. Genetic-engineering techniques also include the use of marker-assisted breeding wherein the genetic "fingerprint" of plants can be used to aid conventional plant breeding. These authors also note the potential for genetic-engineering technology to develop new varieties of crops that could be grown under conditions that reduce some of the adverse environmental impacts of growing food and that contribute to local food production. The rationale parallels the arguments used in discussing the potential of genetic-engineering technology for improving the productive capability of orphan crops in developing countries (Naylor et al., 2004).

The ideological divisions between those who favor and those who oppose the use of GE plants in organic production systems are complex, and in many cases concerns about safety and naturalness are connected to and mask socioeconomic concerns. An example of the complexity was the successful vote in Mendocino County, California, in 2004 to ban the local use of GE organisms in agriculture. The legal focus of the vote was on GE organisms, but it was clear, because of how genetic-engineering technology was linked to issues related to corporate versus local control of agriculture, that the technology was viewed by many of those supporting the measure as a social problem (Walsh-Dilley, 2009). Similarly,

---

[6]A group of growers concerned about the organic purity of an open-pollinated field crop may come together to form a "landscape club," a fee-based organization designed to increase their economic welfare by providing protection against contamination through gene flow from related GE crops. A zone free from genetically modified organisms (GMO-free) would provide similar protection (Jank et al., 2006).

some genetic-engineering proponents argue for including GE products in organic standards and labels at the same time that they argue against the labeling of foods with GE content because they consider GE and non-GE foods to be substantially equivalent products (Klintman, 2002). That position can be understood, in part, as a desire to obtain the economic benefit of some labels while avoiding the cost of being associated with other labels. Those examples underscore the important socioeconomic and sociopolitical dimensions in public debates about genetic-engineering technology. To reconcile those debates over the potential use of genetic engineering in sustainable and developing-country agriculture, it may be wise to heed the suggestion of Ronald and Adamchak (2008) and use various social, environmental, and economic criteria in making decisions on when to use and not to use genetic-engineering technology in agriculture.

CONCLUSIONS

Social dynamics and networks between farmers and within local communities play a substantial role in the decisions that farmers make with respect to the use of GE crops and likely are impacted by the use of and conflicts over those crops. Research on the adoption of other agricultural technologies has demonstrated substantial social impacts on a farm level and a community level. Those impacts include but are not limited to: decreases to and change of composition in the agricultural labor force; better on-farm working conditions; changes in farm and agricultural-industry structure; increases in capital requirements for farmers; and a decline in the socioeconomic viability of some rural communities. Comparable research on the effects of GE crops is lacking, and although it is reasonable to hypothesize that the social impacts of the spread of GE crops have been low due to the assumed scale neutrality of this technology, it is equally reasonable to assume that the social impacts have been numerous and profound. Those questions cannot be answered without short- and long-term empirical research on the social processes surrounding, and the social impacts associated with, the adoption of genetic-engineering technologies at the farm level. Such research must take into account the various contextual factors that are influencing social changes on U.S. farms and rural communities.

Research has demonstrated that farmers' interest in genetic-engineering technology and patterns of adoption are influenced by farmers' social networks and by farmers' associations, private firms, and public actors, including universities. Research also has identified the continuing consolidation of the seed industry and its integration with the chemical industry. The market power of firms that supply seed has not adversely affected farmers' economic welfare so far, but research is needed on how market structure may affect ongoing access to non-GE or single-trait seeds and future seed prices. Furthermore, there has been comparatively little research on how changes in farmer social networks and seed-industry concentration might be affecting farmers' planting decisions and options, overall yield benefits, crop genetic diversity, and economic returns.

A final set of social issues has to do with complex legal issues, including the adoption of and the use of genetic-engineering technology. U.S. and Canadian courts have upheld the legal rights of seed companies to prohibit seed-saving practices through the use of contracts. The issue of gene flow is complicated. One important question being raised is whether adventitious presence of genetic material from GE crops into non-GE crops impinges on the rights of producers, including organic producers, who do not wish to use specific GE traits. The legal debates may mask deeper social and ideological divisions over the use of GE plants and how to define and implement sustainable agricultural practices.

REFERENCES

Anonymous. 2001. Group spells out concerns over genetically modified wheat. CBC-News.ca, August 1. Health section. Available online at http://www.cbc.ca/health/story/2001/07/31/gm_wheat010731.html. Accessed April 1, 2009.

Anonymous. 2008. CAFC again agrees you can't save seed; judge blocks sale of Roche Epo. Patent litigation. *Biotechnology Law Report* 27(3):221–222.

Barham, B.L., J.D. Foltz, D. Jackson-Smith, and S. Moon. 2004. The dynamics of agricultural biotechnology adoption: Lessons from series rBST use in Wisconsin, 1994-2001. *American Journal of Agricultural Economics* 86(1):61–72.

Bauman, D.E. 1992. Bovine somatotropin: Review of an emerging animal technology. *Journal of Dairy Science* 75(12):3432–3451.

Beck, R.L., and H. Gong. 1994. Effect of socioeconomic factors on bovine somatotropin adoption choices. *Journal of Dairy Science* 77(1):333–337.

Berardi, G.M. 1981. Socio-economic consequences of agricultural mechanization in the United States: Needed redirections for mechanization research. *Rural Sociology* 46(3):483–504.

Boehlje, M. 1999. Structural changes in the agricultural industries: How do we measure, analyze and understand them? *American Journal of Agricultural Economics* 81(5):1028–1041.

Bradford, K., J. Alston, and N. Kalaitzandonakes. 2006. Regulation of biotechnology for specialty crops. In *Regulating agricultural biotechnology: Economics and policy.* eds. R.E. Just, J.M. Alston, and D. Zilberman, pp. 683–697. New York: Springer.

Brennan, M.F., C.E. Pray, and A. Courtmanche. 1999. Impact of industry concentration on innovation in the U.S. plant biotech industry. Paper presented at the Transitions in agbiotech: Economics of strategy and policy NE-165 conference (Washington, DC, June 24–25, 1999).

Buttel, F.H., O.F. Larson, and G.W. Gillespie Jr. 1990. *The sociology of agriculture.* New York: Greenwood Press.

Caswell, M.F., K.O. Fuglie, and C.A. Klotz. 1994. Agricultural biotechnology: An economic perspective. Agricultural Economic Report No. 687. U.S. Department of Agriculture–Economic Research Service. Washington, DC.

Chen, L., and F.H. Buttel. 2000. *Dynamics of GMO adoption among Wisconsin farmers*. Madison, WI: Program on Agricultural Technology Studies, University of Wisconsin-Madison and University of Wisconsin-Cooperative Extension.

Cochrane, W.W. 1993. *The development of American agriculture: A historical analysis*. 2nd ed. Minneapolis: University of Minnesota Press.

Conner, D.S. 2003. Pesticides and genetic drift: Alternative property rights scenarios. *Choices: The magazine of food, farm, and resource issues* 1st Quarter: 5–7. Available online at http://www.choicesmagazine.org/2003-1/2003-1-02.pdf. Accessed March 31, 2010.

de Gorter, H., and D. Zilberman. 1990. On the political economy of public good inputs in agriculture. *American Journal of Agricultural Economics* 72(1):131–137.

Delmer, D.P. 2005. Agriculture in the developing world: Connecting innovations in plant research to downstream applications. *Proceedings of the National Academy of Sciences of the United States of America* 102(44):15739–15746.

DuPuis, E.M., and C. Geisler. 1988. Biotechnology and the small farm. *BioScience* 38(6):406–411.

Fernandez-Cornejo, J. 2004. The seed industry in U.S. agriculture: An exploration of data and information on crop seed markets, regulation, industry structure, and research and development. Agriculture Information Bulletin No. 786. U.S. Department of Agriculture–Economic Research Service. Washington, DC. Available online at http://www.ers.usda.gov/publications/aib786/aib786.pdf. Accessed May 26, 2009.

Fernandez-Cornejo, J., and D. Schimmelpfennig. 2004. Have seed industry changes affected research effort? *Amber Waves* 2(1):14–19.

Fernandez-Cornejo, J., and R.E. Just. 2007. Researchability of modern agricultural input markets and growing concentration. *American Journal of Agricultural Economics* 89(5):1269–1275.

Fernandez-Cornejo, J., S. Daberkow, and W.D. McBride. 2001. Decomposing the size effect on the adoption of innovations: Agrobiotechnology and precision farming. Paper presented at the American Agricultural Economic Association Annual Meeting (Chicago, IL, August 5–8, 2001). Available online at http://ageconsearc.umn.edu/bitstream/20527/1/sp01fe02.pdf. Accessed February 23, 2009.

Fuglie, K., N. Ballenger, K. Day, C. Klotz, M. Ollinger, J. Reilly, U. Vasavada, and J. Yee. 1996. Agricultural research and development: Public and private investments under alternative markets and institutions. Agricultural Economics Report No. 735. May. U.S. Department of Agriculture–Economic Research Service. Washington. DC. Available online at http://www.ers.usda.gov/publications/aer735/AER735fm.PDF. Accessed May 31, 2009.

Fulton, M., and K. Giannakas. 2001. Agricultural biotechnology and industry structure. *AgBioForum* 4(2):137–151.

Furtan, W.H., A. Güzel, and A.S. Weseen. 2007. Landscape clubs: Co-existence of genetically modified and organic crops. *Canadian Journal of Agricultural Economics* 55(2):185–195.

Geertson Farms Inc., et al. v. Mike Johanns, et al., and Monsanto Company: Memorandum and order Re: Permanent injunction. 2009. U.S. District Court for the Northern District of California. C 06-01075 CRB, Case#: 3:06-cv-01075-CRB. Decided May 3, 2007. Available online at http://www.aphis.usda.gov/brs/pdf/Alfalfa_Ruling_20070503.pdf. Accessed December 16, 2009.

Goldsmith, P.D. 2001. Innovation, supply chain control, and the welfare of farmers: The economics of genetically modified seeds. *American Behavioral Scientist* (8):1302–1326.

Graff, G.D., and D. Zilberman. 2007. The political economy of intellectual property: Re-examining European policy on plant biotechnology. In *Agricultural biotechnology and intellectual property: Seeds of change*. ed. J.P. Kesan, pp. 244–267. Cambridge, MA: CABI Publishing.

Graff, G.D., S.E. Cullen, K.J. Bradford, D. Zilberman, and A.B. Bennett. 2003. The public-private structure of intellectual property ownership in agricultural biotechnology. *Nature Biotechnology* 21(9):989–995.

Graham, J., and A. Martin. 2004. Biotech wheat pits farmer vs. farmer. *Chicago Tribune*, January 28. p. C-7, News section.

Guehlstorf, N. 2008. Understanding the scope of farmer perceptions of risk: Considering farmer opinions on the use of genetically modified (GM) crops as a stakeholder voice in policy. *Journal of Agricultural and Environmental Ethics* 21(6):541–558.

Hayenga, M.L. 1998. Structural change in the biotech seed and chemical industrial complex. *AgBioForum* 1(2):43–55.

Hill, T. 2009. Personal communication to the Committee on the Impact of Biotechnology on Farm-Level Economics and Sustainability. February 26. Washington, DC.

Jank, B., J. Rath, and H. Gaugitsch. 2006. Co-existence of agricultural production systems. *Trends in Biotechnology* 24(5):198–200.

Jussaume, R.A., Jr., K. Kondoh, and M. Ostrom. 2004. An investigation into the potential introduction of Roundup-Ready wheat. Paper presented at the annual meetings of the Rural Sociological Society (Sacramento, CA, August 15–18, 2004).

Just, D.R., S.A. Wolf, S. Wu, and D. Zilberman. 2002. Consumption of economic information in agriculture. *American Journal of Agricultural Economics* 84(1):39–52.

Kaup, B.Z. 2008. The reflexive producer: The influence of farmer knowledge upon the use of Bt corn. *Rural Sociology* 73(1):62–81.

Kershen, D.L. 2004. Legal liability issues in agricultural biotechnology. *Crop Science* 44(2):456–463.

Klintman, M. 2002. Arguments surrounding organic and genetically modified food labelling: A few comparisons. *Journal of Environmental Policy & Planning* 4(3):247–259.

Kloppenburg, J.A. 1984. The social impacts of biogentetic technology in agriculture: Past and future. In *The social consequences and challenges of new agricultural technologies*. eds. G.M. Berardi and C.C. Geisler, pp. 291–321. Boulder, CO: Westview Press.

Leonard, C. 2006. Monsanto, Pioneer fight for seed market. *The Washington Post*, December 8. Technology section. Available online at http://www.washingtonpost.com/wp-dyn/content/article/2006/12/08/AR2006120800030.html. Accessed November 3, 2009.

Lerner, J., and J. Tirole. 2005. The economics of technology sharing: Open source and beyond. *The Journal of Economic Perspectives* 19(2):99–120.

Lesser, W. 1998. Intellectual property rights and concentration in agricultural biotechnology. *AgBioForum* 1(2):56–61.

Marion, B.W., and R.L. Wills. 1990. A prospective assessment of the impacts of bovine somatotropin: A case study of Wisconsin. *American Journal of Agricultural Economics* 72(2):326–336.

Monsanto. 2010. Academic research agreements. Available online at: http://www.monsanto.com/monsanto_today/for_the_record/academic/research/agreements.asp. Accessed March 22, 2010.

Naylor, R.L., W.P. Falcon, R.M. Goodman, M.M. Jahn, T. Sengooba, H. Tefera, and R.J. Nelson. 2004. Biotechnology in the developing world: A case for increased investments in orphan crops. *Food Policy* 29(1):15–44.

Nytes, A.J., D.K. Combs, G.E. Shook, R.D. Shaver, and R.M. Cleale. 1990. Response to recombinant bovine somatotropin in dairy cows with different genetic merit for milk production. *Journal of Dairy Science* 73(3):784–791.

Phillipson, M. 2001. Agricultural law: Containing the GM revolution. *Biotechnology and Development Monitor* (48):2–5.

Pimentel, D., M.S. Hunter, J.A. Lagro, R.A. Efroymson, J.C. Landers, F.T. Mervis, C.A. McCarthy, and A.E. Boyd. 1989. Benefits and risks of genetic engineering in agriculture. *BioScience* 39(9):606–614.

PIPRA (Public Intellectual Property Resource for Agriculture). 2006. Providing IP support for research consortia: PIPRA forges a novel approach for Pierces' Disease research. *PIPRA*. University of California, Davis, Fall (Issue 6):3.

Pollack, A. 2004. Monsanto shelves plan for modified wheat. *The New York Times*, May 11. p. 1, C section.

———. 2009. Judge rejects approval of biotech sugar beets. *The New York Times*, September 23. p. 3, B section.

Quaintance, H.W. 1984. The influence of farm machinery on production and labor. In *The social consequences and challenges of new agricultural technologies*. eds. G.M. Berardi and C.C. Geisler, pp. 237–248. Boulder, CO: Westview Press.

Rogers, E.M. 2003. *Diffusion of innovations*. New York: Free Press.

Ronald, P.C., and R.W. Adamchak. 2008. *Tomorrow's table: Organic farming, genetics and the future of food*. New York: Oxford University Press.

Ruttan, V.W. 1982. *Agricultural research policy*. Minneapolis: University of Minnesota Press.

Sexton, R.J. 1986. Cooperatives and the forces shaping agricultural marketing. *American Journal of Agricultural Economics* 68(5):1167–1172.

Shi, G., and J.-P. Chavas. 2009. On pricing and vertical organization of differentiated products. Staff Paper No. 535. University of Wisconsin-Madison. Madison, WI. Available online at http://www.aae.wisc.edu/fsrg/. Accessed October 22, 2009.

Shi, G., J.-P. Chavas, and K. Stiegert. 2008. An analysis of bundle pricing: The case of the corn seed market. FSWP2008-01. University of Wisconsin-Madison. Madison, WI. Available online at http://www.aae.wisc.edu/fsrg/. Accessed October 8, 2009.

Squires, G. 2004. Wheat watch. *Wheat Life*. Washington Association of Wheat Growers. Available online at http://www.wawg.org/index.cfm?show=10&mid=59. Accessed August 9, 2009.

Stiegert, K.W., J.P. Chavas, and G. Shi. 2009. Analysis of strategic pricing in the U.S. transgenic cornseed market. Working paper. University of Wisconsin-Madison. Madison, WI.

Thirtle, C., L. Beyers, Y. Ismael, and J. Piesse. 2003. Can GM-technologies help the poor? The impact of Bt cotton in Makhathini Flats, KwaZulu-Natal. *World Development* 31(4):717–732.

Toyama, M., J.W. Heffernan, V.N. Hillers, R.A. Jussaume, Jr., and J.A. Schultz. 2001. Consumers' concerns and behaviors related to biotechnology: Comparison between American and Japanese consumers. Presentation at the IFT Annual Meeting (New Orleans, LA, June 23–27, 2001).

US-EPA (U.S. Environmental Protection Agency). 2008. FIFRA scientific advisory panel: Notice of public meeting. *Federal Register* 73(238):75099–75101.

USDA (U.S. Department of Agriculture). 2010. USDA and DOJ hold first-ever workshop on competition issues in agriculture. Press release No. 0126.10. March 12, 2010. Available online at: http://www.usda.gov/wps/portal/!ut/p/_s.7_0_A/7_0_1OB/.cmd/ad/.ar/sa.retrievecontent/.c/6_2_1UH/.ce/7_2_5JM/.p/5_2_4TQ/.d/2/_th/J_2_9D/_s.7_0_A/7_0_1OB?PC_7_2_5JM_contentid=2010%2F03%2F0126.xml&PC_7_2_5JM_parentnav=LATEST_RELEASES&PC_7_2_5JM_navid=NEWS_RELEASE#7_2_5JM. Accessed March 22, 2010.

Van Es, J.C., D.L. Chicoine, and M.A. Flotow. 1988. Agricultural technologies, farm structure and rural communities in the Corn Belt: Policies and implications for 2000. In *Agriculture and community change in the U.S.: The congressional research reports*. ed. L.E. Swanson, p. 355. Boulder, CO: Westview Press.

Vandenberg, J.M., J.R. Fulton, F.J. Dooley, and P.V. Preckel. 2000. Impact of identity preservation of non-GMO crops on the grain market system. *CAFRI: Current Agriculture, Food and Resource Issues* (01):29–36.

Walsh-Dilley, M. 2009. Localizing control: Mendocino county and the ban on GMOs. *Agriculture and Human Values* 26(1):95–105.

Wilson, W.W. 2009. Personal communication to the Committee on the Impact of Biotechnology on Farm-Level Economics and Sustainability. February 26. Washington, DC.

Wilson, W.W., E.L. Janzen, and B.L. Dahl. 2003. Issues in development and adoption of genetically modified (GM) wheats. *AgBioForum* 6(3):101–112.

Wolf, S.A., D.R. Just, and D. Zilberman. 2001. Between data and decisions: The organization of agricultural economic information systems. *Research Policy* 30(1):121–141.

5

Key Findings, Remaining Challenges, and Future Opportunities

The first generation of genetically engineered (GE) crops has mostly delivered effective pest control for a few major crops because farmers producing these crops spend a lot of time and money on the task, because the firms developing the new seed technologies saw considerable profit potential in doing so, and because adding the traits was relatively straightforward, requiring transformation of a genome at only a single location. The first generation of GE crops continues a reliance on pesticide technology—in-plant toxins or resistance to herbicides—to mitigate pest problems primarily in corn, cotton, and soybean. Thus, the application of genetic-engineering technology to crops has not developed novel means of pest control, such as developing plant mechanisms to resist pest damage, nor has it reached most minor crops.

The next set of challenges for the application of GE-crop technology is to expand to additional crops and to address additional desirable traits, such as drought tolerance, enhanced fertilizer utilization to reduce nutrient runoff, nutritional benefits, renewable energy production, and carbon sequestration. A number of those applications are under development by the private sector, some by the public sector. Clearly, the future agenda for genetic-engineering technology is extensive and of great importance for improvements in agricultural productivity and sustainability in a rapidly changing world.

This chapter opens by summarizing the major findings of our assessment of the farm-level environmental, economic, and social impacts of GE crops. We then identify key remaining challenges that will frame future development and commercialization of genetic-engineering technology in crops. The discussion turns next to the future agenda of GE-crop applications, including general patterns of crop-trait development, implications for future weed-resistance management, and the potential role of GE crops for biofuels. The penultimate section highlights two subjects of research that the committee believes deserve more resources and effort: water quality and social impacts of GE crops. In closing, we discuss options for strengthening public and private research and development to exploit the potential of genetic-engineering technology to contribute more fully to environmental, economic, and social objectives.

KEY FINDINGS

The evidence shows that the planting of GE crops has largely resulted in less adverse or equivalent effects on the farm environment compared with the conventional non-GE systems that GE crops replaced. A key improvement has been the change to pesticide regimens that apply less pesticide or that use pesticides with lower toxicity to the environment but that have more consistent efficacy than conventional pesticide regimens used on non-GE versions of the crops. In the first phase of use, herbicide-resistant (HR) crops have been associated with an increased use of conservation tillage, in particular no-till methods, that can improve water quality and enhance some soil-quality characteristics. That farmers who practice conservation tillage are more likely to adopt GE crops suggests the two technologies are complementary.

At least one potential environmental risk associated with the first phase of GE crops has surfaced: Some adopters of GE crops rely heavily on a single pesticide to control targeted pests, and this leads to a buildup of pest resistance regardless of whether GE crops or non-GE crops are involved. The governmental regulation of GE Bt crops through refuge requirements seems to have proved effective in delaying buildup of insect resistance with two reported exceptions, which have not had major consequences in the United States. Grower decisions to use repeated applications of particular herbicides to some HR crops have led, in some documented cases, to evolved herbicide-resistance problems and shifts in the weed community. In contrast with Bt-crop refuge requirements, no public or private mechanisms for delaying weed resistance have been extensively implemented. If the herbicide-resistance problem is not addressed soon, farmers may increasingly return to herbicides that were used before the adoption of HR crops. Tillage could increase as a pest-management tactic as well. Such actions could limit some of the environmental and personal safety gains associated with the use of HR crops. The newest HR varieties likely will have tolerance to more than one herbicide, and this would allow easier herbicide rotation or mixing, and, in theory, help to improve the durability of herbicide effectiveness. These new stacked varieties will be one more tool to help manage the evolution of weed as well as insect resistance.

The potential for gene flow via cross-pollination between current major GE crops and wild or weedy relatives is limited to cotton in small spatial scales in the United States because the other major GE crops have no native relatives. How this changes in the future will depend on what GE crops are commercialized, whether related species with which they are capable of interbreeding are present, and the consequences of such interbreeding for weed management. Gene flow (i.e., the adventitious presence) of legal GE traits in non-GE crops and derived products remains a serious concern for farmers whose market access depends on adhering to strict non-GE standards. It would appear that the resolution of the issue may require the establishment of enforceable thresholds for the presence of GE material in non-GE crops that do not impose excessive costs on growers and the marketing system.

The literature reviewed in this report indicates that a majority of U.S. farmers who grow soybean, corn, or cotton have generally found GE varieties with herbicide-resistance and insect-resistance traits advantageous because of their superior efficacy in pest control; their concomitant economic, environmental, and presumed personal health advantages; or their convenience. The extent of the benefits varies among locations, crops, and specific genetic-engineering technologies.

After some early evidence of yield disadvantages for some GE varieties in the United States, studies have now shown either a moderate boost in yields of some crops or a neutral yield effect. Some emerging evidence suggests that the attractiveness of genetic-engineering technology for soybean, cotton, and corn has increased the global acreage planted to these crops over what would have been planted otherwise and thereby increased global commodity supplies (World Bank, 2007; Brookes and Barfoot, 2009). Consequently, the adoption of some of the GE crops around the world has put downward pressure on the prices received by U.S. farmers who are growing these crops, holding other factors constant. At the same time, livestock producers and consumers who purchase GE feed and food products may have benefited from the downward price pressure. However, the U.S. and world agricultural economies have been influenced by other factors that tend to increase commodity prices, making empirical verification of the effects difficult.

The economic effects of GE-crop adoption on nonadopters are mixed. To the extent that the use of genetic-engineering technology changes the types and amounts of inputs used, adopters of GE crops can influence the pesticide market. Changes in the price and availability of pesticides affect nonadopters as well. Farmers of non-GE crops in the vicinity of GE-crop farms may experience landscape-level effects from reduced pressure from pests targeted by GE traits. Marketing of non-GE crops may also be affected by GE crops, favorably or unfavorably. For example, products derived from GE and non-GE crops can mix through gene flow or supply contamination. On the other hand, GE crops may create a market premium for non-GE products.

The historic social repercussions of introducing new technologies in agriculture, such as mechanization and the widespread planting of hybrid corn, have been studied extensively, and the results of the studies provide a basis for understanding the general effects of introducing GE varieties of crops. Despite the salience of those effects, however, there has been little investigation of farm-level and community-level social impacts of GE crops. The new seed technologies raise important potential social issues about farm structure, the input and seed choices available to farmers, and the genetic diversity of seeds. Among the known social facts associated with the dissemination of GE crops are the continued consolidation of the seed industry and its integration with the chemical industry. Another is the change in relationships between farmers and their seed suppliers. Testimony to the committee suggested that farmers of major crops have fewer opportunities to purchase non-GE seed of the best-yielding cultivars even when a GE trait is not perceived to be required in a particular cropping situation. As genetic-engineering technology matures and moves into its next phase, it is imperative that the full array of social issues involved be identified and investigated in depth.

REMAINING CHALLENGES FACING GENETICALLY ENGINEERED CROPS

Potential crop-biotechnology developments stir discussion around five issues. The treatment and resolution of those issues hold implications for long-term sustainability for farmers, including both adopters and nonadopters of GE crops.

First, the success of genetic-engineering technology in the United States has altered the seed industry by spurring consolidation of firms and integration with the chemical industry. Those developments continue to alter seed and pest-control options in the market, expanding pest-control options for some farmers and possibly limiting them for others, including those who do not grow GE crops. The resulting concentrated and legally enforceable private control of plant genetic material contrasts sharply with public plant-breeding programs that traditionally have fostered public access to discoveries, especially if GE varieties are developed for crops from which farmers have traditionally saved seed for the next year's crop (such as wheat, barley, and oat). Although corporate consolidation may offer greater economies of scale, it also is accompanied by the possibility of less competition and higher seed expenses, which may lower farmer returns, reduce pest-control options, and limit the benefits of commercialization of genetic-engineering technology to a few widely grown crops. There is not yet clear evidence of a lack of competition among the large companies that produce GE-crop technologies, but these trends should be monitored, and their effects ameliorated to remedy social losses that result.

Second, how the intensive use of current and prospective GE organisms will directly affect the natural environment differently from other agricultural production systems is incompletely understood (Ervin et al., 2003). Relatively few studies have provided integrated assessments of the indirect effects of GE crops on pest damage to non-GE crops and on the full suite of ecosystem services on the landscape scale. For example, the concurrent effects on regional water quality of shifting tillage and pesticide regimens with the introduction of GE crops on regional water quality conditions remain poorly documented and understood. Knowledge of the spatial and temporal effects on ecological health—favorable or unfavorable—assumes greater importance as the evolution of herbicide resistance in weeds alters patterns of herbicide use to make up for the loss of glyphosate's efficacy on some species, and as novel GE plants, such as those for energy and nonfood uses, approach commercialization. Evaluation and monitoring of the ecological health of soils, water quality and quantity, and air quality will provide the information needed for developing the most productive yet sustainable agricultural systems for the future (Ervin et al., 2003).

Third, progress in developing GE varieties for most "minor" crops (e.g., fruits and vegetables) and for other "public goods" purposes not served well by private markets has been slow. Minor crops play important roles in the agricultural sectors of many states. Some minor crops, such as sunflowers and grain sorghum, have been considered to be poor candidates for the HR traits because of the existence of near-relative weeds (grain sorghum) or native ancestral populations (sunflowers). However, that risk does not explain the relative dearth of research and development (R&D) on GE varieties of minor crops as a whole, especially fruits and vegetables. The high fixed investment, patent protection involving GE traits, and the regulatory expense of commercializing GE seeds have lessened the ability of small companies and public-sector crop-breeding programs to develop commercially viable GE varieties of most minor crops. It seems clear that more effort should be devoted to the enhancement of minor crops with the best available genetic-engineering techniques even though the commercial rewards per crop may be small. Private and public

R&D programs for GE crops have not yet led to the commercialization of many additional plant traits, such as improved tolerance to drought and increased fertilizer-use efficiency that decreases nutrient runoff, a contributor to nonpoint-source water pollution. This is not to suggest that GE seeds represent a "silver bullet" technology to solve all of these problems. Such GE trait developments may or may not turn out to be the most cost-effective approach to solving these issues, but exploration is necessary to evaluate their relative efficacy. Even though the basic technology is over 20 years old, agricultural biotechnology in some regards is still in its infancy. The rate of progress of genetic-engineering technology for some purposes suggests that more time and resources and new institutional relationships, such as public–private R&D collaborations, are desirable for the technology to reach its potential. Traits that have not yet received much attention, such as improved nutrient absorption and enhanced food value of crops, should be emphasized in the future so that we can gain the knowledge needed for weighing how the advances of genetic-engineering technology could present options for addressing all aspects of food supply, energy security, and environmental challenges.

Fourth, the presence of transgenic material in non-GE products should be addressed. The current definition of organic food in the United States excludes the use of GE materials in production and handling processes, so organic farmers must take steps to ensure that their production methods do not expose their crops to GE traits. To avail themselves of market premiums for certified organic crops, they must incur costs for keeping their products separate from GE crops. Food producers who market products as non-GE face similar challenges to prevent co-mingling of GE and non-GE crops during storage and distribution. In those ways, the introduction of GE crops influences the decisions and operations of farmers and food producers who do not use the technology.

Fifth, U.S. farmers who grow GE crops may face market restrictions from some countries or retail firms on the importation or sale of the crops or products made from the crops. Under some international agreements, some countries impose these restrictions because of perceived food safety or environmental risks or for other reasons. Assuming that those actions satisfy international treaty rules, such market-access restrictions to some extent slow the development of a global market for GE crops. One effect of the trade restrictions has been to limit the market demand for GE crops.

The potential of GE-crop varieties to address the world's emerging food-supply, energy, and environmental problems hinges on how those challenges are resolved. Success in resolving them may allow genetic-engineering technology to become even more transformational in fostering sustainable agricultural systems for farmers. This important agenda frames the discussion of future GE-crop applications.

FUTURE APPLICATIONS OF GENETICALLY ENGINEERED CROPS

In addition to expanding existing GE traits into other crops, the reach of genetic-engineering technology could be extended through the development and commercialization of new traits. Traits beyond those designed to control pests could have substantial benefits in fields other than agriculture, such as food and energy security. This section summarizes the present pattern of R&D of novel GE traits in the private and public sectors and highlights areas in which new traits could be especially useful to improving agricultural sustainability.

Patterns of Genetically Engineered Products in Development

The GE crops now being planted by U.S. farmers were developed over long gestation periods. Given that fact, the current portfolio of GE crops does not wholly capture the range of new crops being readied for commercialization that could cover the U.S. farm and global landscape in the future. For instance, companies are developing crops with more intricate pest-control mechanisms (see Box 5-1), but they are also engineering

---

BOX 5-1
New Traits Reduce Refuge Requirement and
Introduce Second Mode of Herbicide Resistance

On July 20, 2009, the Environmental Protection Agency (EPA) announced final approval for commercialization of SmartStax™ corn hybrids. These hybrids produce multiple proteins for the control of lepidopteran pests and for the control of corn rootworm (two of these proteins must be used together to induce significant rootworm mortality). The hybrids also have transgenes conferring resistance to glyphosate and glufosinate. Because of the multiple toxins affecting single pests, this group of hybrids was approved by EPA with a refuge requirement of 5 percent in the Corn Belt and 20 percent in the Cotton Belt (US-EPA, 2009). Those refuge requirements are reduced from the original requirements of 20 percent in the Corn Belt and 50 percent in the Cotton Belt for previous transgenic Bt corn hybrids. Resistance to the different broad-spectrum herbicides may help in delaying the evolution of resistance in weeds in areas where glyphosate-resistant populations have not been identified. However, growers must assume appropriate stewardship for these herbicides in order to manage the future evolution of herbicide-resistant weed populations.

traits that improve crop tolerance to stress or that provide benefits directly to consumers. Recent research also indicates that novel forms of pest control may be part of the next generation of GE crops. As an example, Baum et al. (2007) report positive laboratory findings that ribonucleic-acid (RNA) interference technology, a plant-based method for pest management, results in larval stunting and mortality in several coleopteran species controlled by Bt crops. Predicting future crop biotechnologies is somewhat difficult because companies, for reasons of competitiveness and patent protection, may not fully share the specifics of planned releases. However, three recent global surveys of product-quality innovations in the development phase of genetic-engineering technology help in discerning general trends in GE-crop development and, in particular, why quality-improving innovations from genetic-engineering technology have not yet been more numerous (Graff et al., 2009).

The combined findings of the surveys show that less than 5 percent of the innovations reach the regulatory and commercialization phases. As might be expected, R&D activity has been uneven among different trait categories because of variation in the difficulty of achieving selected transformations and in the economic value of the traits. For example, traits governing content and composition of macronutrients—proteins, oils, and carbohydrates—and traits that control fruit ripening have more readily reached later stages of R&D, and few products with enhanced micronutrients, functional food components, or novel esthetics have approached commercialization (Graff et al., 2009). The analysts concluded that product-quality innovation appears more responsive to demand in intermediate markets for processing and feed attributes than to demand in retail markets for improved or novel products. They also noted that many of the observed traits offer potential efficiency gains in agriculture and improvements in natural-resource systems and a potential to reduce environmental impacts both because they will decrease input requirements and because they will reduce adverse externalities of crop production, processing, or consumption. For example, new traits may increase the efficiency of livestock-feed digestibility and by so doing increase the value of the feed to farmers. Traits that improve nitrogen-use efficiency that are on the horizon will bring value to farmers and could contribute to reducing agriculture's effects on water quality. The potential for environmental improvement from such traits depends on the degree to which they improve input-use efficiency and the extent to which farmers expand production because of lower unit costs.

The rate of product-quality innovations in genetic-engineering technology identified in the surveys decreased considerably after 1998. The authors noted that the cause of the decline remains conjectural. It coincided with the decrease in the number of transgenic field trials conducted in the United States and Europe, with the exit from the market of a number of small biotechnology companies that were not deeply involved in the first generation of GE pest-control crops, and with changes in the regulatory environment in Europe that restricted the introduction of GE varieties there.

Data on field tests of genetic-engineering innovations corroborate the slowdown. A critical part of new variety development is field testing to ensure that the desired traits will perform under commercial production conditions and that no important environmental risks are associated with release of a GE organism. The release of new GE varieties or organisms into the environment is regulated, in part, through field-release permits monitored by the U.S. Department of Agriculture (USDA) Animal and Plant Health Inspection Service (APHIS) (Fernandez-Cornejo and Caswell, 2006).[1] The overall number of field releases of plant varieties for testing purposes is a useful indicator of R&D efforts in GE crops. By the end of 2008, about 15,000 applications had been received by APHIS since 1987, and nearly 14,000 (93 percent) had been approved (ISB and USDA-APHIS, 2009). Annual applications peaked in 1998 with 1,206, and annual approvals peaked in 2002 with 1,141 (Figure 5-1). Most applications approved for field testing in 1987–2008 involved major crops, particularly corn with 6,648 applications approved, followed by soybean (1,554), cotton (912), potato (817), tomato (637), wheat (413), alfalfa (385), and tobacco (363). Applications approved during this time included GE varieties with herbicide resistance (25.1 percent), insect resistance (20.1 percent), improved product quality (flavor, appearance, or nutrition) (18.2 percent), and agronomic properties, such as drought resistance (13.4 percent) (Figure 5-2).

After sufficient field testing, an applicant may petition APHIS for a determination of "nonregulated" status to facilitate commercialization of a product. If, after extensive review, APHIS determines that unconfined release does not pose a substantial risk to agriculture or the environment, the organism is "deregulated" and can be moved and planted without APHIS authorization (Fernandez-Cornejo and Caswell, 2006). Petitions for deregulated varieties peaked in 1995–1997 with 14-15 petitions per year and have been below 10 petitions every year thereafter (ISB and USDA-APHIS, 2009). As of June 5, 2009, APHIS had received 119 petitions for deregulation and had approved 76. The deregulated varieties had HR traits (38 percent), IR traits (28 percent), product-quality traits (15 per-

---

[1] If a plant is engineered to produce a substance that "prevents, destroys, repels, or mitigates a pest", it is considered to be a pesticide and is subject to regulation by the Environmental Protection Agency (Fernandez-Cornejo and Caswell, 2006). Thus, all Bt crops are included.

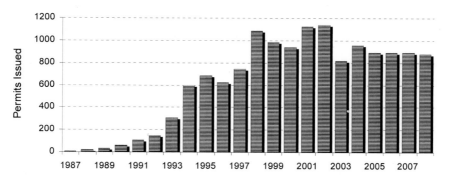

FIGURE 5-1 Number of permits for release of genetically engineered varieties approved by APHIS.
SOURCE: ISB and USDA-APHIS, 2009.

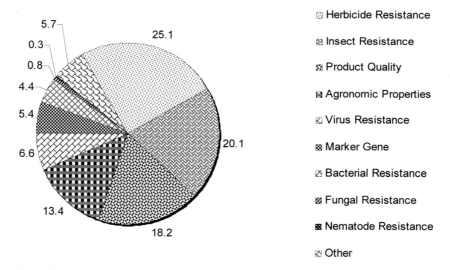

FIGURE 5-2 Approved field releases of plant varieties for testing purposes by trait (percent).
SOURCE: ISB and USDA-APHIS, 2009.

cent), virus-resistance traits (11 percent), and agronomic traits (6 percent) (ISB and USDA-APHIS, 2009).

Another study of the APHIS GE-crop release data investigated the private or commercial and public-goods aspects of traits being tested (Welsh and Glenna, 2006). The authors analyzed releases in 1993–2002 to answer two research questions:

- "To what extent have universities mimicked the for-profit sector in agricultural biotechnology by focusing their transgenic research on a relatively few proven genotypes (traits)?"
- "To what extent have universities mimicked the for-profit sector in agricultural biotechnology by focusing their transgenic research on the relatively few major (commercially dominant) agronomic crops?"

They categorized the crop releases into major traits—herbicide resistance, insect resistance, and product quality—and "other" GE traits. Product quality includes the alteration of a particular crop's characteristics to make it more valuable to food, feed, or energy manufacturing firms, such as a higher lysine concentration in corn that would be useful to livestock producers. Other GE traits were nematode resistance, fungus resistance, bacterium resistance, virus resistance, and agronomic properties (such as yield). They also categorized the releases by whether they were major crops (soybean, corn, wheat, alfalfa, and cotton) or minor crops (such as other field crops, vegetables, and fruits planted on smaller acreages).

The results indicated that the research profiles of universities in 1993–2002 were less dominated by the major traits than were the profiles of for-profit firms. A similar pattern emerged for major crops: About 71 percent of notices filed with APHIS by for-profit firms entailed research on major crops compared with 32.6 percent filed by universities. Moreover, work on minor crops differed among universities depending on their region (for example, apples in the Northeast and citrus in the Southeast). The authors examined whether the relationships changed during three periods: 1993–1995, 1996–1998, and 1999–2002. The research profiles of the for-profit firms remained fairly uniform, but universities looked more like for-profit firms in the later periods. That trend was especially pronounced for major traits: Almost 73 percent of notices filed by universities entailed at least one major trait in 1999–2002 compared with around 35 percent in earlier periods. The proportion of research on major crops by universities also increased.

To probe those relationships in more depth, Welsh and Glenna constructed a commercial index for each release. The index was computed by assigning scores of 1 for research on a major crop and 1 for research on a major trait. Research on a minor crop was scored as –1 and research on a minor trait was scored as –1. Possible index values were therefore 2, 1, 0, –1, and –2. A higher index value indicated greater research emphasis on more commercially relevant crops and traits, and a lower value indicated more emphasis on crops and traits with smaller markets. Plotting the value of the index for universities and private-sector firms over time revealed that universities and private-sector firms showed increasingly similar research trajectories, which emphasized major crops and major traits in genetic-engineering technology. The time-series relationship was also found to be statistically significant (Welsh and Glenna, 2006).

Results of those three studies of GE-trait developments indicated that some GE products in various phases of development serve purposes other than pest-control traits dominant today, but they have not been commercialized. The reasons for this vary with the crop and the trait, but include a small anticipated market, lack of access to technology, uncertainty about consumer acceptance, potential spillover to weedy relatives or gene transfer to non-GE cultivars, and high regulatory costs. During the process of innovation, commercialization, and adoption, all organizations weigh the costs and benefits of their decisions (Bradford et al., 2006). It is not surprising that the GE crops that were commercialized first provided substantial net benefits to the innovators, the seed companies, and adopting farmers.

Implications of Genetically Engineered Crops for Weed Management

An unmet challenge for GE crops documented in Chapter 2 revolves around sustaining the efficacy of particular herbicides. Maintaining their efficacy holds important implications for future farm economics and environmental sustainability. A key concept in understanding herbicide-resistance effects in weeds is the open, unregulated access of all farmers to the common pool of pest susceptibility. The presence of this condition leads to individual decisions that may impose user costs on the whole population of farmers, to suboptimal overall management, and to increasing total social costs (Hardin, 1968).

The data show that GE-crop cultivars resistant to glyphosate dominate in the corn, soybean, and cotton production regions across the United States. Results of a six-state survey to assess crop rotations by growers of GE crops showed that rotation of a GE crop with a non-GE crop was most commonly followed by rotation of a GE crop with a GE crop (Shaw et al., 2009). Rotation of a GE with a non-GE crop was more common in the Midwest than in the South. With the high acreage of glyphosate-resistant crops being planted, substantial changes in herbicide use occurred; notably, fewer herbicides were used (Young, 2006); the most common current herbicide tactic reported is one to three applications of glyphosate (Givens et al., 2009). A related change in production practice attributable to the adoption of HR crops that has implications for weed management was the increased adoption of conservation tillage (Givens et al., 2009). The overall grower assessment of the effect of HR crops on weed population densities was that there were fewer weeds because of the use of glyphosate on HR crops (Kruger et al., 2009). Few growers felt that weed populations would shift to species that had evolved resistance to glyphosate, understood the importance of alternative tactics to control weeds, or perceived the role of selection pressure caused by the use of glyphosate on HR crops (Johnson et al., 2009). Given that changes in weed communities in response to production practices have been a consistent problem in agriculture, future weed challenges are likely to increase quickly (Baker, 1991; Owen, 2001; Heard et al., 2003a, 2003b). To summarize, growers valued the immediate benefits of weed control without appreciating the long-term risks attributable to the tactics they used. That behavioral response might be expected given many farmers' desire to meet short-run financial needs and the fact that other growers may not take similar control actions.

Recent survey data show that growers emphasize the convenience and simplicity of the glyphosate-based cropping systems while discounting the importance of diversification of weed-management practices (Owen, 2008b). That prevailing attitude has several likely results: Given growers' apparent unfamiliarity with or lack of concern about the implications of selection pressure for weed communities, weed shifts are inevitable; and growers' short-term interest in killing weeds rather than managing them will result in the loss of crop yield and farm profits in the long run unless innovation in weed-control technology occurs. The efficacy of glyphosate on a broad spectrum of weeds, its ability to kill larger weeds, and the fact that glyphosate can be applied to HR crops almost regardless of crop stage of growth all reinforce growers' perception of the simplicity and convenience of glyphosate-based programs.

Weed management implies that there is a critical period of weed control (CPWC) when weed interference must be eliminated to protect crop-yield potential (Nieto et al., 1968; Kasasian and Seeyave, 1969; Swanton and Weise, 1991; Kneževič et al., 2002, 2003). The CPWC is the same whether the crop is GE or non-GE. There is considerable information about the proper timing of glyphosate applications that will provide protection of yield potential and the related economic loss of crop yield in response to untimely glyphosate application (Gower et al., 2003; Dalley et al., 2004; Cox et al., 2005, 2006; Stahl, 2007). Many growers may recognize the relationship between early weed interference with HR crops and the resulting economic loss to profitability or may face high individual cost or risk in changing their behavior. The perception of success with glyphosate results in the repeated use of this herbicide without consideration of alternative strategies. Recurrent application of any herbicide will cause shifts in the weed community that support the evolving dominance of weeds that are not susceptible. Growers must use diversified weed-management practices, recognize the importance of understanding the biology of the cropping system, and give appropriate consideration to more sustainable weed-management programs (Kneževič et al., 2003; Owen and Zelaya,

2005; Owen, 2008a). Furthermore, unless growers collectively adopt more diverse weed-management practices, individual farmer's actions will fail to delay herbicide resistance to glyphosate because the resistant genes in weeds easily cross farm boundaries. Some form of private or public collective action may avert this classic management of the commons problem in which individual actions to apply glyphosate have adverse spillover effects on the entire community of farmers (Hardin, 1968). Further research that results in new HR traits and other efficient means of weed control likely will lessen this problem.

Potential for Biofuels

Amid diminishing reserves of fossil fuels, heightened concern about climate change, and growing demand for domestic energy production, biofuels have emerged as an important supplementary fuel that may have considerable potential in supplying future energy needs. First-generation biofuels are serving as fuel extenders, displacing only a small percentage of gasoline consumption in the United States (Energy Information Administration, 2007). Nevertheless, because some existing biofuel crops can be produced in a manner that reduces greenhouse-gas emissions relative to fossil fuels and because they reduce dependence on volatile oil import markets, governments around the world have supported production of biofuel crops with subsidies and mandates (Koplow, 2007). However, although, as noted, some biofuel technologies can reduce greenhouse-gas emissions (Farrell et al., 2006), that might not always be the case, depending on production practices (Fargione et al., 2008; Searchinger et al., 2008). Furthermore, when they increase the demand for food crops, biofuel production can contribute to higher prices and shortages of food like those witnessed in 2008 (Tyner and Taheripour, 2008; Sexton et al., 2009).

In the United States, ethanol is produced principally from corn, approximately 80 percent of which is grown with GE varieties. Similarly, biodiesel in the United States is produced almost entirely from soybean; about 92 percent U.S. acres in soybean produce GE varieties. If those GE crops have increased yields, they may reduce biofuel costs.

If developed appropriately, new GE-crop technologies could play an important role in ameliorating further the adverse economic and environmental impacts of biofuels (Sexton et al., 2009). Indeed, some governments, like that of the United States, have tailored their biofuels policy to include, among other instruments, some support for next-generation technologies to overcome existing limitations (e.g., Rajagopal et al., 2007; Rajagopal and Zilberman, 2008). For example, genetic-engineering technology may help to improve the agronomic characteristics of plants used for cellulosic ethanol, which are not yet widely commercialized. Whereas ethanol is produced today only from the starch in plants, developments in microbiology allow the cellulosic material to be converted to biofuel. Genetic-engineering technology holds the potential to generate higher yields of these crops and to improve the amount of liquid fuel obtainable per plant by altering plants' genetic code in beneficial ways. It may provide these benefits without environmental damage. Because many of the plants that may provide fuel in the future have not been commercially farmed, it may be possible to improve plant genetics to maximize their energy-yield potential, minimize the costs of converting cellulosic plant material to liquid fuel, and devise best-management practices or mitigation for the environment.

RESEARCH PRIORITIES RELATED TO GENETICALLY ENGINEERED CROPS

Water-Quality Monitoring and Evaluation

Nonpoint pollution is the leading cause of water-quality impairments across the United States, extending into ocean estuaries, bays, and gulfs (US-EPA, 2007). Agriculture remains the largest source of these nonpoint pollution flows by volume, and much of the pollution stems from cropland operations. The predominant contaminants include sediment from land erosion and nutrient and pesticide residues not used or retained for growing crops. For example, a recent analysis estimated that agriculture contributes 70 percent of the nitrogen and phosphorus that enters the Gulf of Mexico and that corn production accounts for the majority of the nitrogen and corn and soybean production account for one-fourth of the phosphorus (Alexander et al., 2008). Particularly in view of the huge dead zone that has formed in the Gulf of Mexico, lessening such pollution has high national priority.

As explained in Chapter 2, evidence has begun to emerge that GE crops are often associated with changes in cropping practices that should lead to an improvement in the nation's water quality. The changes include shifts to conservation tillage or no-till techniques that leave more residues on the cropland surface and thereby reduce water runoff that contains sediment, nutrient, and pesticide contaminants. They also include the use of pesticides, such as glyphosate, that are less toxic and more quickly degrading than conventional crop herbicides and insecticides. The latter effects could also reduce contamination of groundwater and wells on farms from spills and mixing operations.

Because monitoring and research resources have been inadequate, those potential water-quality impacts of GE crops have not been documented. The committee received testimony from the U.S. Geological Survey that explained the lack of comprehensive data and analysis that could identify and estimate the magnitude of the potential improvements in water quality (Gilliom and Meyer, personal communication). Those effects are among the largest environmental effects of agriculture, and we recommend that more resources be devoted to the important tasks of spatial and temporal monitoring of how agricultural practices influence water quality. Monitoring changes associated with the adoption of GE cultivars is important given that the rapid, widespread adoption of those cultivars may have large impacts on water quality by changing agricultural practices. The resulting information could influence the design of future environmental policies and agriculture programs. Such critical intelligence would help to improve the efficacy and cost-effectiveness of achieving regional and national water-quality standards, and thereby improve farm sustainability.

Social Issues in the Use of Genetically Engineered Crops

Accumulated research in the social sciences has verified that the processes of technological development and diffusion do not take place in a social vacuum. Choices made by those who create new technologies and decisions made by others regarding whether to use the technologies are influenced by political, economic, and sociocultural factors. The social impacts of the use of the technologies are influenced by the same factors. A particular technology can have dissimilar social impacts depending on the context within which it is adopted. It is reasonable to hypothesize, on the basis of the existing body of knowledge, that the adoption of particular genetic-engineering technologies has a variety of social, economic, and political impacts, and that these impacts would not be the same at all times and in all regions and cultures.

As noted in Chapter 4, the amount of research on the social processes and effects associated with the development and use of GE crops has been inadequate and has not matched what took place previously in the cases of agricultural mechanization or even the use of bovine somatotropin in dairy production. Thus, there is little empirical evidence to which the present committee can point to that delineates the full array of social impacts of the adoption of GE-crop technology. That includes a lack of research on the social impact on farmers of companies legally enforcing their intellectual-property rights in GE seeds. Research that has been conducted on the role of industry–farmer social networks in influencing the development of new seed technologies, legal issues related to GE-crop technology, resistance to the use of the technology in organic production systems, and the development of the open-source breeding movement does suggest, however, that genetic-engineering technology is socially contested by some groups. Those findings and trends reinforce the need for research on the social processes associated with genetic-engineering technology, including its social impacts, to inform the decision-making processes of technology developers, farmers, and policy makers. The committee recommends the development of this research agenda, which should lead to findings important for addressing the sociocultural issues that will arise in connection with broader adoption of GE crops in the future.

ADVANCING POTENTIAL BENEFITS OF GENETICALLY ENGINEERED CROPS BY STRENGTHENING COOPERATION BETWEEN PUBLIC AND PRIVATE RESEARCH AND DEVELOPMENT

The rapid adoption by U.S. farmers of the first generation of GE soybean, cotton, and corn varieties illustrates the speed and scope with which agricultural systems can be improved if appropriate products and systems are available. This report documents how GE varieties contribute to the sustainability of agriculture related to the production of those major crops. Expanding the effects to additional crops and further improving the technology will require an expansive program of R&D. Private companies are already working to develop additional traits that will improve the productivity and sustainability of agriculture in the United States and worldwide. However, both the private and public sectors must play vigorous, if often times different, roles if the full potential of genetic-engineering technology to foster a more sustainable agriculture is to be realized. In developing analogous traits for other crops, such as GE varieties of "minor" crops and additional GE traits to meet broader public environmental and social objectives (e.g., improved water quality and carbon sequestration), the active involvement of universities and nongovernment institutions will be crucial given the flexibility of such institutions in selecting research objectives when funding is available. Developing the most appropriate agenda for such research will require extensive stakeholder involvement, including input from adopters and nonadopters of GE crops, environmental and social interest groups, and industry representatives.

Public investment and innovation in genetic-engineering technology includes several phases, such as discovery, scaling up of innovations, regulatory research, commercialization, production, and marketing. Life-science innovations increasingly take place within the educational–industrial complex, where research universities and public research institutions are engaged mainly in the early stages of innovation, and startup and major corporations are engaged more in product development (Graff et al., 2009). Much of the academic research addresses basic problems with uncertain outcomes that may result in new commercial innovations or in knowledge that has either pure public-good properties in the economic sense, such as basic-research discoveries that are nonrival and nonexclusive (Just et al., 2008), or properties that are not easy to appropriate, so revenues cannot be collected. Thus, such academic research will receive underinvestment by the private sector and warrants public-sector support (Dasgupta and David, 1994; Dasgupta, 1999).

Not all discoveries from academic research in genetic-engineering technology result in nonrival and nonexcludable products. For example, some scientists patent their innovations; these are then excludable products that can be licensed to other scientists, nonprofit organizations, or industry. A national survey in 2004 revealed that about 25 percent of responding scientists had filed for a patent since 2000. About 15 percent of the scientists had been issued a patent, and just under 8 percent had licensed their invention for use by private or public parties (Buccola et al., 2009). The responding scientists expressed slight support for patenting compared to the belief that publicly supported scientists should focus on knowledge with nonexcludable benefits (a mean of 2 versus a mean of 5 on a 6-point scale) (Buccola et al., 2009). Some scientists may be inclined to patent their discoveries in case they contribute to future, commercially developed technology.

In those cases where university discoveries and innovations are basic proofs of concepts, the private sector often undertakes the development effort, upscaling manufacturing capacity and commercializing the technology. Companies that invest in development of university innovations frequently buy rights to university patents to secure intellectual-property protection and monopoly power once the products are developed. Without that protection, the firms may be unwilling to invest the capital necessary to move the technology to the commercialization stage.

There is likely to be underinvestment in commercialization of biotechnology innovation by the private sector because companies with implicit monopoly rights associated with patents aim to maximize their own profit without taking full account of consumers' welfare gains that result from the lower prices associated with an innovation. Firms cannot capture the potential benefits external to farmers and consumers, such as reduction in downstream pollution. The underinvestment in research makes a case for public-sector development of GE varieties as long as social benefits exceed social costs or social-equity objectives defined by elected representatives are achieved (de Gorter and Zilberman, 1990; Just et al., 2008). The situation suggests that public-sector research should emphasize the development of genetic-engineering technologies in specialty crops and innovation of other kinds, such as traits that may lead to a reduction in greenhouse-gas emissions from crop or livestock production or novel varieties that conserve water resources. The ability of the public and private sectors to develop new gene technologies depends on the costs of innovation, which may include access to intellectual-property rights and regulatory requirements. A tailored and targeted regulatory approach to GE-crop trait development and commercialization that meets human and environmental safety standards while minimizing unnecessary expenses could enhance progress on this front (Ervin and Welsh, 2006).

In reinforcement of those conceptual points, recent studies of genetic-engineering R&D have concluded that publicly funded research programs can complement private-sector R&D efforts in developing the full potential of agricultural biotechnology (Graff and Zilberman, 2001; Glenna et al., 2007; Buccola et al., 2009). There are several reasons for that conclusion. First, federal and state support encourages more basic research, whereas industry and foundations support more applied research in U.S. universities (Buccola et al., 2009). Downstream (i.e., more applied) research tends to be legally and economically more excludable than upstream (i.e., more basic) research. Publicly funded research offers the highest potential for achieving public goods, such as the basic science of genetic mechanisms, broadly accessible platform technologies, and nonmarket environmental services (Buccola et al., 2009). Second, industry collaborations with academic scientists have affirmed the necessity of a strong independent university research sector in helping to provide credible evaluations of new technology (Glenna et al., 2007). Third, both publicly and privately supported research assist in the transfer of basic discoveries in genetic-engineering technology, such as plant-genome characterizations, into useful crop-plant applications (Graff and Zilberman, 2001). For example, publicly supported basic research on *Arabidopsis thaliana* has provided an enormous store of information on basic plant biology, which in turn has enhanced our ability to produce commercially important advances in crops of all kinds. New institutional mechanisms are needed to provide an uninterrupted "pipeline" from basic research to field application (Graff et al., 2003). Fourth, commercialization of orphan and minor crops requires a special role for public R&D because the improvement of such crops often will not lead to sufficient profit to attract private-sector investment, even though the crops are important to many farmers and consumers. Fifth, public funding of academic research and government research fosters investigation that is too risky, or not sufficiently profitable, to be attractive to the private sector.

The evidence assembled in this report makes clear that the first generation of GE soybean, cotton, and corn varieties has generally been economically and environmentally advantageous for U.S. farmers who have adopted the technologies. The next generation of genetic-engineering technologies being reported by industry suggests that it is intent on enhancing those benefits and going beyond to new traits, such as drought and heat tolerance and enhanced fertilizer utilization that may indirectly reduce nutrient runoff, and new applicability to minor crops, renewable energy, climate change, and nutritional qualities. The public sector must complement industry by developing genetic-engineering technologies for crops that have insufficient markets to justify R&D and regulatory expense and to develop socially valuable public-goods applications. We envision the research and technology agenda to include individual private and public activities as well as private–public collaborations and to encompass work on the three essential components of sustainable development: environmental, social, and economic. Furthermore, we recommend that this agenda be undertaken in a program of research, technology development, and education that maximizes the potential synergies between the two sectors and their strengths and that limits redundancies and tradeoffs. Such an integrated approach would have universities, government, and nonprofit organizations leading in the development of traits that deliver public goods, including basic discoveries and such environmental issues as improved regional water quality. The private sector would continue to lead in the commercialization of GE crops for which there are adequate market incentives.

REFERENCES

Alexander, R.B., R.A. Smith, G.E. Schwarz, E.W. Boyer, J.V. Nolan, and J.W. Brakebill. 2008. Differences in phosphorus and nitrogen delivery to the Gulf of Mexico from the Mississippi river basin. *Environmental Science & Technology* 42(3):822–830.

Baker, H.G. 1991. The continuing evolution of weeds. *Economic Botany* 45:445–449.

Baum, J.A., T. Bogaert, W. Clinton, G.R. Heck, P. Feldmann, O. Ilagan, S. Johnson, G. Plaetinck, T. Munyikwa, M. Pleau, T. Vaughn, and J. Roberts. 2007. Control of coleopteran insect pests through RNA interference. *Nature Biotechnology* 25(11):1322–1326.

Bradford, K., J. Alston, and N. Kalaitzandonakes. 2006. Regulation of biotechnology for specialty crops. In *Regulating agricultural biotechnology: Economics and policy*. eds. R.E. Just, J.M. Alston, and D. Zilberman, pp. 683–697. New York: Springer.

Brookes, G., and P. Barfoot. 2009. Global impact of biotech crops: Income and production effects, 1996-2007. *AgBioForum* 12(2):184–208.

Buccola, S., D. Ervin, and H. Yang. 2009. Research choice and finance in university bioscience. *Southern Economic Journal* 75(4):1238–1255.

Cox, W.J., R.R. Hahn, P.J. Stachowski, and J.H. Cherney. 2005. Weed interference and glyphosate timing affect corn forage yield and quality. *Agronomy Journal* 97(3):847–853.

Cox, W.J., R.R. Hahn, and P.J. Stachowski. 2006. Time of weed removal with glyphosate affects corn growth and yield components. *Agronomy Journal* 98(2):349–353.

Dalley, C.D., J.J. Kells, and K.A. Renner. 2004. Effect of glyphosate application timing and row spacing on weed growth in corn (*Zea mays*) and soybean (*Glycine max*). *Weed Technology* 18(1):177–182.

Dasgupta, P. 1999. Science as an institution: Setting priorities in a new socio-economic context. Paper presented at the Plenary session on science in society at the UNESCO/ICSU world conference on science (Budapest, Hungary, June 26–July 1, 1999). Available online at http://www.econ.cam.ac.uk/faculty/dasgupta/science.pdf. Accessed September 16, 2009.

Dasgupta, P., and P.A. David. 1994. Toward a new economics of science. *Research Policy* 23(5):487–521.

de Gorter, H., and D. Zilberman. 1990. On the political economy of public good inputs in agriculture. *American Journal of Agricultural Economics* 72(1):131–137.

Energy Information Administration. 2007. Biofuels in the U.S. transportation sector. U.S. Department of Energy. Washington, DC., Available online at http://www.eia.doe.gov/oiaf/analysispaper/biomass.html. Accessed January 12, 2010.

Ervin, D., and R. Welsh. 2006. Environmental effects of genetically modified crops: Differentiated risk assessment and management. In *Regulating agricultural biotechnology: Economics and policy*. eds. R.E. Just, J.M. Alston and D. Zilberman, pp. 301–326. New York: Springer.

Ervin, D.E., R. Welsh, S.S. Batie, and C.L. Carpentier. 2003. Towards an ecological systems approach in public research for environmental regulation of transgenic crops. *Agriculture, Ecosystems and Environment* 99(1–3):1–14.

Fargione, J., J. Hill, D. Tilman, S. Polasky, and P. Hawthorne. 2008. Land clearing and the biofuel carbon debt. *Science* 319(5867):1235–1238.

Farrell, A.E., R.J. Plevin, B.T. Turner, A.D. Jones, M. O'Hare, and D.M. Kammen. 2006. Ethanol can contribute to energy and environmental goals. *Science* 311(5760):506–508.

Fernandez-Cornejo, J., and M.F. Caswell. 2006. The first decade of genetically engineered crops in the United States. Economic Information Bulletin No. 11. April. U.S. Department of Agriculture–Economic Research Service. Washington, D.C. Available online at http://www.ers.usda.gov/publications/eib11/eib11.pdf. Accessed June 15, 2009.

Gilliom, R.J., and M.T. Meyer. 2009. Personal communication to the Committee on the Impact of Biotechnology on Farm-Level Economics and Sustainability. February 26. Washington, DC.

Givens, W.A., D.R. Shaw, W.G. Johnson, S.C. Weller, B.G. Young, R.G. Wilson, M.D.K. Owen, and D. Jordan. 2009. A grower survey of herbicide use patterns in glyphosate-resistant cropping systems. *Weed Technology* 23(1):156–161.

Glenna, L.L., W.B. Lacy, R. Welsh, and D. Biscotti. 2007. University administrators, agricultural biotechnology, and academic capitalism: Defining the public good to promote university-industry relationships. *Sociological Quarterly* 48(1):141–163.

Gower, S.A., M.M. Loux, J. Cardina, S.K. Harrison, P.L. Sprankle, N.J. Probst, T.T. Bauman, W. Bugg, W.S. Curran, R.S. Currie, R.G. Harvey, W.G. Johnson, J.J. Kells, M.D.K. Owen, D.L. Regehr, C.H. Slack, M. Spaur, C.L. Sprague, M. VanGessel, and B.G. Young. 2003. Effect of postemergence glyphosate application timing on weed control and grain yield in glyphosate-resistant corn: Results of a 2-yr multistate study. *Weed Technology* 17(4):821–828.

Graff, G., and D. Zilberman. 2001. An intellectual property clearinghouse for agricultural biotechnology. *Nature Biotechnology* 19(12):1179–1180.

Graff, G., D. Zilberman, and A. Bennett. 2009. The contraction of agbiotech product quality innovation. *Nature Biotechnology* 27(8):702–704.

Graff, G.D., S.E. Cullen, K.J. Bradford, D. Zilberman, and A.B. Bennett. 2003. The public-private structure of intellectual property ownership in agricultural biotechnology. *Nature Biotechnology* 21(9):989–995.

Hardin, G. 1968. The tragedy of the commons. *Science* 162(3859):1243–1248.

Heard, M.S., C. Hawes, G.T. Champion, S.J. Clark, L.G. Firbank, A.J. Haughton, A.M. Parish, J.N. Perry, P. Rothery, R.J. Scott, M.P. Skellern, G.R. Squire, and M.O. Hill. 2003a. Weeds in fields with contrasting conventional and genetically modified herbicide-tolerant crops. I. Effects on abundance and diversity. *Philosophical Transactions of the Royal Society of London B* 358:1819–1832.

Heard, M.S., C. Hawes, G.T. Champion, S.J. Clark, L.G. Firbank, A.J. Haughton, A.M. Parish, J.N. Perry, P. Rothery, D.B. Roy, R.J. Scott, M.P. Skellern, G.R. Squire, and M.O. Hill. 2003b. Weeds in fields with contrasting conventional and genetically modified herbicide-tolerant crops. II. Effects on individual species. *Philosophical Transactions of the Royal Society of London B* 358:1833–1846.

ISB (Information Systems for Biotechnology), and USDA-APHIS (U.S. Department of Agriculture–Animal and Plant Health Inspection Service). 2009. Field test release applications in the U.S. Available online at http://www.isb.vt.edu/cfdocs/fieldtests1.cfm. Accessed June 5, 2009.

Johnson, W.G., M.D.K. Owen, G.R. Kruger, B.G. Young, D.R. Shaw, R.G. Wilson, J.W. Wilcut, D.L. Jordan, and S.C. Weller. 2009. U.S. farmer awareness of glyphosate-resistant weeds and resistance management strategies. *Weed Technology* 23(2):308–312.

Just, R.E., D.L. Hueth, and A. Schmitz. 2008. *Applied welfare economics*. Northampton, MA: Edward Elgar Publishing.

Kasasian, L., and J. Seeyave. 1969. Critical periods for weed competition. *International Journal of Pest Management: Part A* 15(2):208–212.

Kneževič, S.Z., S.P. Evans, E.E. Blankenship, R.C. Van Acker, and J.L. Lindquist. 2002. Critical period for weed control: The concept and data analysis. *Weed Science* 50(6):773–776.

Kneževič, S.Z., S.P. Evans, and M. Mainz. 2003. Row spacing influences the critical timing for weed removal in soybean (*Glycine max*). *Weed Technology* 17(4):666–673.

Koplow, D.N. 2007. Biofuels, at what cost? Government support for ethanol and biodiesel in the United States: 2007 update. Earth Track, Inc. Cambridge, MA.

Kruger, G.R., W.G. Johnson, S.C. Weller, M.D.K. Owen, D.R. Shaw, J.W. Wilcut, D.L. Jordan, R.G. Wilson, M.L. Bernards, and B.G. Young. 2009. U.S. grower views on problematic weeds and changes in weed pressure in glyphosate-resistant corn, cotton, and soybean cropping systems. *Weed Technology* 23(1):162–166.

Nieto, H.J., M.A. Brondo, and J.T. Gonzales. 1968. Critical periods of the crop growth cycle for competition from weeds. *PANS: Pest Articles & News Summaries* 14:159–166.

Owen, M.D.K. 2001. Importance of weed population shifts and herbicide resistance in the Midwest USA Corn Belt. In *Pesticide behaviour in soils and water: Proceedings of a symposium organised by the British Crop Protection Council*, pp. 407–412 (Brighton, UK, November 12–15, 2001). Farnham, Surrey, UK: British Crop Protection Council.

———. 2008a. Weed species shifts in glyphosate-resistant crops. *Pest Management Science* 64(4):377–387.

———. 2008b. Glyphosate resistant crops and evolved glyphosate resistant weeds—the need for stewardship. In *Proceedings of the 5th International Weed Science Congress*, p. 51 (Vancouver, Canada, June 23–27, 2008). ed. K. Hurle. International Weed Science Society.

Owen, M.D.K., and I.A. Zelaya. 2005. Herbicide-resistant crops and weed resistance to herbicides. *Pest Management Science* 61(3):301–311.

Rajagopal, D., and D. Zilberman. 2008. Environmental, economic and policy aspects of biofuels. *Foundations and Trends in Microeconomics* 4(5):353–468.

Rajagopal, D., S.E. Sexton, D. Roland-Holst, and D. Zilberman. 2007. Challenge of biofuel: Filling the tank without emptying the stomach? *Environmental Research Letters* 2(4).

Searchinger, T., R. Heimlich, R.A. Houghton, F. Dong, A. Elobeid, J. Fabiosa, S. Tokgoz, D. Hayes, and T.-H. Yu. 2008. Use of U.S. croplands for biofuels increases greenhouse gases through emissions from land-use change. *Science* 319(5867):1238–1240.

Sexton, S., D. Zilberman, D. Rajagopal, and G. Hochman. 2009. The role of biotechnology in a sustainable biofuel future. *AgBioForum* 12(1):130–140.

Shaw, D.R., W.A. Givens, L.A. Farno, P.D. Gerard, D. Jordan, W.G. Johnson, S.C. Weller, B.G. Young, R.G. Wilson, and M.D.K. Owen. 2009. Using a grower survey to assess the benefits and challenges of glyphosate-resistant cropping systems for weed management in U.S. corn, cotton, and soybean. *Weed Technology* 23(1):134–149.

Stahl, L. 2007. Application timing is key in optimizing returns. *News and Information*, May 14. Available online at http://www.extension.umn.edu/extensionnews/2005/applicationtiming07.html. Accessed July 8, 2009.

Swanton, C.J., and S.F. Weise. 1991. Integrated weed management: The rationale and approach. *Weed Technology* 5(3):657–663.

Tyner, W., and F. Taheripour. 2008. Biofuels, policy options, and their implications: Analyses using partial and general equilibrium approaches. *Journal of Agricultural and Food Industrial Organization* 6(2).

US-EPA (U.S. Environmental Protection Agency). 2007. National water quality inventory: Report to Congress; 2002 reporting cycle. October 2007. EPA 841-R-07-001. Washington, DC. Available online at http://www.epa.gov/305b/2002report/report2002305b.pdf. Accessed July 29, 2009.

———. 2009. Pesticide fact sheet: SmartStax™. Washington, DC. Available online at http://www.epa.gov/oppbppd1/biopesticides/pips/smartstax-factsheet.pdf. Accessed September 28, 2009.

Welsh, R., and L.L. Glenna. 2006. Considering the role of the university in conducting research on agri-biotechnologies. *Social Studies of Science* 36(6):929–942.

World Bank. 2007. *World development report 2008: Agriculture for development*. Washington, DC: The World Bank.

Young, B.G. 2006. Changes in herbicide use patterns and production practices resulting from glyphosate-resistant crops. *Weed Technology* 20(2):301–307.

Appendix A

Herbicide Selection

Herbicide[a]	Classification[b]
Haloxyfop (aryloxyphenoxypropionates) Clethodim (cyclohexanediones) Pinoxaden (phenylpyrazolines)	ACCase inhibitors (A[(1)])
Alachlor (chloroacetanilides) Diphenamid (acetamides) Flufenacet (oxyacetamides) Fentrazamide (tetrazolinones)	Mitosis inhibitors (K_3[(15)])
Bensulfuron-methyl (sulfonylureas) Imazethapyr (imidazolinones) Cloransulam-methyl (triazolopyrimidines)	ALS inhibitors (B[(2)])
Atrazine (symmetrical triazines)	Photosynthesis II inhibitors (C_1[(5)])
Dicamba (benzoic acids)	Synthetic auxins (O[(4)])
Glufosinate (organophosphorus)	Glutamine synthetase inhibitor (H[(10)])
Glyphosate (organophosphorus)	Enolpyruvyl shikimate-3-phosphate (EPSP) synthase inhibitor (G[(9)])

*Reference dose.
†Lethal dose.
§Lethal concentration.

Mode of Action[c]	Relative Toxicity
Inhibitor of acetyl coenzyme-A carboxylase (ACCase), a pivotal enzyme in plant fatty acid biosynthesis.	Varies based on specific chemical compound; likely to be carcinogenic to humans according to EPA *Proposed EPA Weight-of-the-Evidence Categories* (US-EPA, 2009a).
Inhibition of cell division (long-chain fatty acid inhibitor).	Slightly toxic; oral RfD* of 1×10^{-2} mg/kg-day. Critical effects: hemosiderosis, hemolytic anemia (US-EPA, 2009b).
Inhibition of the acetolactate synthase (ALS) enzyme resulting in cessation of the biosynthesis of essential branched chain amino acids (leucine, valine, and isoleucine).	Varied; generally low acute and chronic toxicity to humans and not likely to be carcinogenic.
Inhibit photosynthesis by binding with a specific protein in the photosystem II complex.	Slightly to moderately toxic; oral RfD of 3.5×10^{-2} mg/kg-day. Critical effects: decreased body weight gain (other effect: cardiac toxicity and moderate-to-severe dilation of the right atrium (EXTOXNET, 2009a; US-EPA, 2009c)).
The specific mode of action is not well defined, however these herbicides mimic the endogenous auxin indoleacetic acid, a plant hormone that stimulates growth and appears to negatively affect cell wall plasticity and nucleic acid metabolism.	Relatively nontoxic. Acute oral LD_{50}[†] in rats is 1707 mg/kg, dermal LC_{50}[§] in rabbits is >2000 mg/kg. However, eye irritation in rabbit is extreme.
Inhibits the activity of glutamine synthetase, which causes ammonia buildup in the cell. The ammonia destroys cell membranes.	Practically nontoxic by ingestion; some increase in absolute and relative kidney weights in males. No observed carcinogenesis. Oral RfD of 4×10^{-4} mg/kg-day (NPIC, 2009; US-EPA, 2009d).
Inhibits the EPSP synthase, which leads to depletion of essential aromatic amino acids (tryptophan, tyrosine, and phenylalanine).	Generally nontoxic. Increased incidence of renal tubular dilation in third generation weanlings. Oral RfD of 1×10^{-1} mg/kg-day (US-EPA, 2009e).

continued

Herbicide[a]	Classification[b]
Amides, pyridiazinones, pyridines, isoxazoles, pyrazoles, triazoles, triketones, and others	Carotenoid biosynthesis inhibitors
Fluridone (unclassified)	$(F_1^{(12)})$
Mesotrione (triketones)	$(F_2^{(28)})$
Amitrole (triazoles)	$(F_3^{(11)})$
Metribuzin (triazinones, asymmetrical triazines)	Photosynthesis II inhibitors $(C_1^{(5)})$
Dicarboximide herbicides, triazolone herbicides, diphenylethers, N-phenylphthalimides, oxadiazoles, thiadiazoles, triazolinones, and others	Protoporphyrinogen Oxidase (PPO) inhibitors $(E^{(14)})$

Mode of Action[c]	Relative Toxicity
Inhibit the catabolic degradation of tyrosine to plastoquinones (important for photosynthesis and carotenoid biosynthesis) and tocopherol (vitamin E, which protects biological membranes against oxidative stress and the photosynthetic apparatus against photo-inactivation).	Varied; developmental toxicity, probable human carcinogen, phytotoxic (US-EPA, 2004, 2006, 2009f, 2009g).
Inhibition of phytoene desaturase, an enzyme essential for carotenoid biosynthesis.	
Inhibition of hydroxyphenylpyruvate dioxygenase (HPPD), an enzyme involved in the synthesis of plastoquinone (PQ) and tocopherol (vitamin E).	
Target site is generally considered unknown with the exception of amitrole, which inhibits lycopene cyclase and clomazone that is reported to inhibit an early step in the nonmevalonic acid isoprenoid pathway ultimately leading to carotenoid synthesis.	
Inhibit photosynthesis by binding with a specific protein in the photosystem II complex.	A slightly toxic compound in EPA toxicity class III. Liver and kidney effects, decreased body weight, mortality; no indications of carcinogenic effects. RfD of 2.5×10^{-2} mg/kg-day (EXTOXNET, 2009b; US-EPA, 2009h).
Inhibits the protoporphyrinogen oxidase (PPO) which is in the chlorophyll synthesis pathway. The PPO inhibition starts a reaction in the cell that ultimately causes the destruction of cell membranes. The leaking cell membranes rapidly dry and disintegrate.	Varied; including compounds having: slight toxicity (EPA category III); classified as a "not likely" human carcinogen to "likely to be a human carcinogen" (US-EPA, 2001).

continued

Herbicide[a]	Classification[b]
2,4-D (phenoxyacetic herbicide)	Synthetic auxins ($O^{(4)}$)

[a]Herbicide listed is one example of the herbicide family.
[b]Herbicide mode of action code according to the Herbicide Resistance Action Committee (HRAC, 2010) and the Weed Science Society of America (WSSA) classification (Senseman and Armbrust, 2007). The capitalized letter is the HRAC classification and the superscript number is the WSSA classification.
[c]The indicated herbicide mode of action is from the HRAC and WSSA descriptions as well as other citations.

REFERENCES

EXTOXNET (Extension Toxicology Network). 2009a. Atrazine. Corvallis, OR: Oregon State University. Available online at http://extoxnet.orst.edu/pips/atrazine.htm. Accessed November 14, 2009.

———. 2009b. Metribuzin. Corvallis, OR: Oregon State University. Available online at http://extoxnet.orst.edu/pips/metribuz.htm. Accessed November 14, 2009.

HRAC (Herbicide Resistance Action Committee). 2010. Classification of herbicides according to mode of action. Washington, DC. Available online at http://www.hracglobal.com/Publications/ClassificationofHerbicideModeofAction/tabid/222/Default.aspx. Accessed January 21, 2010.

NPIC (National Pesticide Information Center). 2009. Glufosinate. Corvallis, OR: Oregon State University. Available online at http://extoxnet.orst.edu/pips/atrazine.htm. Accessed November 14, 2009.

Senseman, S.A., and K. Armbrust, eds. 2007. *Herbicide handbook*. 9th ed., p. 458. Lawrence, KS: Weed Science Society of America.

US-EPA (U.S. Environmental Protection Agency). 2001. Pesticide fact sheet: Flumioxazin. 7501C. Office of Prevention, Pesticides, and Toxic Substances. Washington, DC. Available online at http://www.epa.gov/opprd001/factsheets/flumioxazin.pdf. Accessed January 21, 2010.

———. 2004. Report of the Food Quality Protection Act (FQPA) Tolerance Reassessment Progress and Risk Management Decision (TRED) for Fluridone. 7508C. Office of Prevention, Pesticides, and Toxic Substances. Washington, DC. Available online at www.epa.gov/oppsrrd1/REDs/fluridone_tred.pdf. Accessed January 21, 2010.

———. 2006. Pesticide fact sheet: 1,2,4-triazole, triazole alanine, triazole acetic acid: Human health aggregate risk assessment in support of reregistration and registration actions for triazole-derivative fungicide compounds. Office of Prevention, Pesticides, and Toxic Substances. Washington, DC. Available online at www.epa.gov/opprd001/factsheets/tetraHHRA.pdf. Accessed January 21, 2010.

———. 2009a. User's manual for RSEI, version 2.2.0 [1996 - 2006 TRI data]. Office of Prevention, Pesticides, and Toxic Substances. Washington, DC. Available online at http://www.epa.gov/oppt/rsei/pubs/RSEI%20Users%20Manual%20V2.2.0.pdf. Accessed June 24, 2009.

Mode of Action[c]	Relative Toxicity
Mimics a plant growth regulator. The specific mode of action is not well defined, however these herbicides mimic the endogenous auxin indoleacetic acid, a plant hormone that stimulates growth and appears to negatively affect cell wall plasticity and nucleic acid metabolism.	Critical effects include: hematologic, hepatic, and renal toxicity. RfD of 1×10^{-2} mg/kg-day.

————. 2009b. Alachlor (CASRN 15972-60-8). National Center for Environmental Assessment. Washington, DC. Available online at http://www.epa.gov/ncea/iris/subst/0129.htm. Accessed January 21, 2010.

————. 2009c. Atrazine (CASRN 1912-24-9). National Center for Environmental Assessment. Washington, DC. Available online at http://www.epa.gov/ncea/iris/subst/0209.htm. Accessed January 21, 2010.

————. 2009d. Glufosinate-ammonium (CASRN 77182-82-2). National Center for Environmental Assessment. Washington, DC. Available online at http://www.epa.gov/ncea/iris/subst/0247.htm. Accessed January 21, 2010.

————. 2009e. Glyphosate (CASRN 1071-83-6). National Center for Environmental Assessment. Washington, DC. Available online at http://www.epa.gov/ncea/iris/subst/0057.htm. Accessed January 21, 2010.

————. 2009f. Pesticide fact sheet. Office of Prevention, Pesticides, and Toxic Substances. Washington, DC. Available online at http://www.epa.gov/ncea/iris/subst/0057.htm. Accessed January 21, 2010.

————. 2009g. Tembotrione chemical documents. Office of Prevention, Pesticides, and Toxic Substances. Washington, DC. Available online at http://www.epa.gov/opprd001/fact-sheets/tembotrione.htm. Accessed January 21, 2010.

————. 2009h. Metribuzin (CASRN 21087-64-9). Office of Prevention, Pesticides, and Toxic Substances. Washington, DC. Available online at http://www.epa.gov/NCEA/iris/subst/0075.htm. Accessed January 21, 2010.

Appendix B

Tillage Systems

Below is an outline of general tillage and weed-management practices for corn, soybean, and cotton. Tillage and weed-control practices, however, vary greatly across regions of the United States and within a region based upon grower preference, soil texture, structure, and erosion potential for each individual farm.

Conventional or Conservation Tillage

Pretillage – may include shredding of cornstalks, usually in the fall, shortly after harvest.

Primary Tillage – types of equipment include moldboard plow, chisel plow, field cultivator, and tandem disk depending upon preference by the grower and soil erosion potential. For example, moldboard plow would be avoided if the fields are designated as Highly Erodible Land. Consequently, the chisel plow, field cultivator, or tandem disk are primary tillage implements in conservation tillage systems. Primary tillage can occur in the fall or the spring, depending on soil conditions and erosion potential of the soil. Furthermore, primary tillage may be done as a zone over the row where crops will ultimately be planted (strip tillage).

Secondary Tillage – types of equipment include tandem disk, field cultivator, chisel disk, disk harrow, harrow, and other implements. The type of equipment used depends on grower preferences, machinery complement, the region of the country, soil structure and texture, soil erosion potential, and environmental conditions during the season. Secondary tillage operations typically occur in the spring. Secondary tillage can involve multiple passes but quite often a series of implements are pulled in tandem, making it a one-pass operation. If growers are practicing conservation tillage, more than 30 percent of residue is present on the soil surface after planting.

Weed management – may or may not include a half rate or full rate of soil-applied herbicides with residual activity before or after planting (preemergence or early postemergence to the crop). Modern sprayers are typically 90 feet in width or more, and this step requires very little time and fuel use. Some growers, however, have tanks mounted on their planters and spray the residual herbicides at planting, thus saving an operation. Herbicide-resistant crops are typically treated with glyphosate postemergence to the crop and weeds and may include other herbicides with different mechanisms of action. A second (and possibly a third) application of glyphosate is typically applied, especially in cotton and soybean. If one or more weeds have become resistant to glyphosate or the farmer's herbicide of choice, then the grower may apply a substitute herbicide or use a tillage operation to control the weed.

Organic growers substitute tillage (cultivation) operations, typically three or more, for herbicides for weed control. The first operation is usually performed just after corn or soybean has emerged, using a rotary hoe. This step is usually followed by two cultivations, the first occurring when corn or soybean is quite small.

No-Till

Pretillage – some growers may shred cornstalks.

Weed management – some growers will use a nonselective "burn-down" herbicide, such as glyphosate, to kill existing weeds if they are present before planting. Farmers may also wait until after planting for the initial herbicide application. This treatment may include a combination of herbicides that provide residual control of weeds that emerge later in the season. Postemergence applications of herbicides may be used later in the season depending on the weed infestation.

Planting – involves more sophisticated versions of the planters used for conventional tillage. Typically, the planters have heavy coulters and other attachments to clear the residue from the previous crop in the planting row. Not all soils are suitable for no-till, especially wet and heavy soils in northern latitudes, and no-till may lead to increased pest occurrence because conventional tillage may reduce insect, pathogen, and weed occurrence. On the other hand, no-till is well adapted to well-drained soils in warm regions because no-till improves soil-water infiltration and reduces soil evaporation, thereby providing more soil water to the crop.

Appendix C

Biographical Sketches of Committee Members

David E. Ervin (*Chair*) is professor of environmental management, professor of economics, and fellow at the Center for Sustainable Processes and Practices at Portland State University. Dr. Ervin also serves on the board of the United States Society for Ecological Economics. He teaches economics of sustainability, business environmental management, and global environmental issues. His research and writing work includes university–industry research relationships in agricultural biotechnology, risk management of transgenic crops, voluntary business environmental management, and green technology. He recently directed a multi-university and multidisciplinary research project on public goods and university–industry relationships in agricultural biotechnology funded by the U.S. Department of Agriculture (USDA). He holds a B.S. and an M.S. from the Ohio State University and a Ph.D. in agricultural and resource economics from Oregon State University.

Yves Carrière is a professor of insect ecology in the Department of Entomology at the University of Arizona. He is an expert on the interactions between insects and transgenic plants, environmental impacts of transgenic crops, and integrated pest management. He is an associate editor of the *Journal of Insect Science*. Dr. Carrière received a B.Sc. and an M. Sc. in biology from Laval University and holds a Ph.D. in entomology and behavioral ecology from Simon Fraser University.

William J. Cox, a professor of crop science, joined the Cornell University faculty on an extension–research appointment in 1984. He has served in several capacities, including department associate chairman and extension leader. He recently evaluated the effects of transgenic seed on the yield and economics of corn production. His research also focuses on the environmental, biotic, and management interactions that influence the growth, development, yield, and quality of corn, soybeans, and wheat. He collaborates closely with soil scientists, animal scientists, plant pathologists, entomologists, and plant breeders in an effort to quantify whole-plant physiological responses of the crop to the environmental, biotic, and crop management interactions. He is a senior associate editor of the *Agronomy Journal* and the electronic publication *Crop Management*. Dr. Cox holds a Ph.D. in crop science from Oregon State University. He received an M.S. in agronomy from California State University-Fresno and a B.S. in history from the College of the Holy Cross.

Jorge Fernandez-Cornejo is an agricultural economist in the Resource and Rural Economics Division of U.S. Department of Agriculture Economic Research Service (ERS). He currently works on the adoption and diffusion of agricultural technologies, agricultural biotechnology, and economics of biofuel production. Since joining ERS in 1990, Dr. Fernandez-Cornejo has researched U.S. farmers' experience with biotechnology in the first decade of its adoption and the effects of the technology on farmers' decision-making process. He has also studied the seed industry. He has a Ph.D. in operations research and agricultural economics and a master's in chemical engineering from the University of Delaware, an M.A. in energy and resources from the University of California, Berkeley, and a B.S. in industrial engineering. Dr. Fernandez-Cornejo has expertise in agricultural economics, farm management, integrated pest management, and farm-level impacts of transgenic seed.

Raymond A. Jussaume, Jr., is professor and chair of the Department of Community and Rural Sociology at Washington State University. His academic appointment includes teaching, extension–outreach, and research. The main thrust of his research has been to contribute to a growing international research agenda on the globalization of agri-food systems and various strategies for improving agricultural sustainability. Most recently, much of his research has been focused on how agricultural sustainability can be enhanced by increasing the extent to which agri-food systems are "localized". He recently published several journal articles evaluating Washington State farmers' attitudes toward biotechnology. Dr. Jussaume was a participant at the National Research Council's Conference on Incorporating Science, Economics, and Sociology in Developing Sanitary and

Phytosanitary Standards in International Trade. He received his Ph.D. in development sociology from Cornell University.

Michele C. Marra is a professor of agricultural economics at North Carolina State University and an extension specialist. A production economist, she has concentrated on economic issues surrounding integrated pest management and the characteristics of agricultural innovations that affect farmer choice. She works on the farm-level impacts of crop biotechnologies and the economics of precision farming. Her recent publications have analyzed the benefits of and risks posed by adopting new agricultural technologies and the effects of agricultural biotechnology on farmer welfare. Dr. Marra is a member of the American Agricultural Economics Association and served as the associate editor of the *American Journal of Agricultural Economics* from 2004 to 2007. She has a Ph.D. in economics and an M.S. and a B.S. in agricultural economics from North Carolina State University.

Micheal D.K. Owen has a Ph.D. in agronomy and weed science from the University of Illinois. He is associate chair in the Department of Agronomy of Iowa State University. He has extensive expertise in weed dynamics, integrated pest management, and crop risk management. His objective in extension programming is to develop information about weed biology, ecology, and herbicides that can be used by growers to manage weeds with cost efficiency and environmental sensitivity. His work is focused on supporting management systems that emphasize a combination of alternative strategies and conventional technology. Dr. Owen has published extensively on farm-level attitudes toward transgenic crops and their impacts; selection pressure, herbicide resistance, and other weed life-history traits; and tillage practices.

Peter H. Raven is the president of the Missouri Botanical Garden; a George Engelmann Professor of Botany, Washington University, St. Louis; adjunct professor of biology, University of Missouri, St. Louis and St. Louis University; and a member of the National Academy of Sciences (NAS). He earned his Ph.D. at the University of California, Los Angeles. Dr. Raven was a member of President Clinton's Council of Advisors on Science and Technology. He served for 12 years as home secretary of NAS and is a member of the academies of science of Argentina, Brazil, China, Denmark, India, Italy, Mexico, Russia, Sweden, and the U.K. Dr. Raven's primary research interests are in the systematics, evolution, and biogeography of the plant family Onagraceae; plant biogeography, particularly in the tropics and Southern Hemisphere; and tropical floristics and conservation. The author of numerous books and reports, both popular and scientific, Dr. Raven was a coauthor of *Biology of Plants* and *Environment*.

L. LaReesa Wolfenbarger conducts research on ecological effects of transgenic crops and agricultural practices and on land management for grassland bird conservation at the University of Nebraska, Omaha. Other research interests include the effects of agriculture on grassland ecosystems and the ecology of grassland ecosystems in agricultural landscapes. She has published several articles on the relationship between genetically engineered organisms and the environment and on the ecological risks and benefits related to genetically engineered plants. Her research also seeks to understand the responses of avian communities and reproduction to habitat variation and to management practices on restored grasslands, remnant prairies, and marginal agricultural habitats. Her other work includes synthesizing science on agricultural biotechnology, chairing a committee for a departmental graduate student program, organizing public symposia on environmental issues, and managing a 160-acre prairie preserve. Dr. Wolfenbarger earned her Ph.D. in ecology from Cornell University.

David Zilberman has been a professor in the Department of Agricultural and Resource Economics of the University of California, Berkeley, since 1979. His research interests are in agricultural and nutritional policy, economics of technological change, economics of natural resources, and microeconomic theory. He is a fellow of the American Agricultural Economics Association and the Association of Environmental and Resource Economists, which have recognized many of his publications on the adoption and regulation of agricultural biotechnology for their quality and value to the field. He received his B.A. in economics and statistics from Tel Aviv University in Israel and his Ph.D. in agricultural and resource economics from the University of California, Berkeley. Dr. Zilberman has expertise in the intersection of biotechnology and politics and economics and agricultural marketing. He has recently published on biofuels and biotechnology marketing.